T0362104

Finite Element Analysis and
DESIGN OF METAL
STRUCTURES

Finite Element Analysis and
DESIGN OF METAL STRUCTURES

EHAB ELLOBODY

RAN FENG

BEN YOUNG

AMSTERDAM • BOSTON • HEIDELBERG • LONDON
NEW YORK • OXFORD • PARIS • SAN DIEGO
SAN FRANCISCO • SINGAPORE • SYDNEY • TOKYO
Butterworth-Heinemann is an imprint of Elsevier

Butterworth-Heinemann is an imprint of Elsevier
225 Wyman Street, Waltham, MA 02451, USA
The Boulevard, Langford Lane, Kidlington, Oxford OX5 1GB, UK

Library of Congress Cataloging-in-Publication Data
A catalog record for this book is available from the Library of Congress.

British Library Cataloguing-in-Publication Data
A catalogue record for this book is available from the British Library.

ISBN: 978-0-12-416561-8

For information on all Butterworth-Heinemann publications
Visit our Web site at www.books.elsevier.com

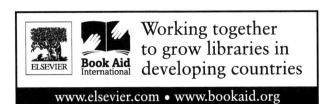

Working together
to grow libraries in
developing countries

www.elsevier.com • www.bookaid.org

CONTENTS

CHAPTER 1

Introduction

1.1. GENERAL REMARKS

Most of finite element books available in the literature, e.g. Refs [1.1−1.7], deal with explanation of finite element method as a widely used numerical technique for solving problems in engineering and mathematical physics. The books mentioned in Refs [1.1−1.7] were written to provide basic learning tools for students in civil and mechanical engineering. The aforementioned books highlighted the general principles of finite element method and the application of method to solve practical problems. Numerous books are also available in the literature, as examples in Refs [1.8−1.26], addressing the behavior and design of metal structures. The books mentioned in Refs [1.8−1.26] have detailed the analysis and design of metal structural elements considering different design approaches. However, up-to-date, there is a dearth in the books that detail and highlight the implementation of finite element method in analyzing metal structures. Extensive numerical investigations using finite element method were presented in the literature as research papers on metal columns, beams, beam columns, and connections. However, detailed books that discuss the general steps of finite element method specifically as a complete work on metal structures and connections are rarely found in the literature, leading to the work presented in this book.

There are many problems and issues associated with modeling of metal structures in the literature that students, researchers, designers, and academics need to address. This book provides a collective material for the use of finite element method in understanding the behavior and structural performance of metal structures. Current design rules and specifications of metal structures are mainly based on experimental investigations, which are costly and time consuming. Hence, extensive numerical investigations were performed in the literature to generate more data, fill in the gaps, and compensate the lack of data. This book also highlights the use of finite element methods to improve and propose more accurate design guides for metal structures, which is rarely found in the literature. The book contains examples for finite element models developed for

Finite Element Analysis and Design of Metal Structures
DOI: http://dx.doi.org/10.1016/B978-0-12-416561-8.00001-9

different metal structures as well as worked design examples for metal structures. The authors hope that this book will provide the necessary material for all interested researchers in the field of metal structures. The book can also act as a useful teaching tool and help beginners in the field of finite element analysis of metal structures. The book can provide a robust approach for finite element analysis of metal structures that can be understood by undergraduate and postgraduate students.

The book consists of eight well-designed chapters covering necessary topics related to finite element analysis and design of metal structures. Chapter 1 provides a general background for the types of metal structures, mainly on columns, beams, and tubular connections. The three topics present the main structural components that form any metal frame, building, or construction. Detailing the analysis of these components would enable understanding the overall structural behavior of different metal structures. The chapter also gives a brief review of the role of experimental investigations as the basis for finite element analysis. Finally, the chapter highlights the importance of finite element modeling and current design codes for understanding the structural performance of metal structures.

Chapter 2 provides a simplified review of general steps of finite element analysis of metal structures. The chapter enables beginners to understand the fundamentals of finite element analysis and modeling of complicated structural behavior of metals. The chapter also includes how to divide a metal structural element into finite elements and how to select the best type of finite elements to represent the overall structural element. The chapter provides a brief review of the selection of displacement functions and definition of strain—displacement and stress—strain relationships. In addition, Chapter 2 also presents a brief review of the formation of element stiffness matrices and equations, the assemblage of these equations, and how the assembled equations are solved for unknowns.

Chapter 3 focuses on finite element modeling of metal structures and details the choice of element type and mesh size that can accurately simulate the complicated behavior of different metal structural elements. The chapter details how the nonlinear material behavior can be efficiently modeled and how the initial local and overall geometric imperfections were incorporated in the finite element analysis. Chapter 3 also details modeling of different loading and boundary conditions commonly applied to metal structures. The chapter focuses on the finite element modeling using any software or finite element package, as an example in this book, the use of ABAQUS [1.27] software in finite element modeling.

Chapter 4 extends the information covered in Chapter 3 to explain and detail the commonly used linear and nonlinear analyses in finite element modeling of metal structures. The chapter also explains the analyses generally used in any software and details as an example the linear and nonlinear analyses used by ABAQUS [1.27]. The chapter also contains a brief survey and background of the linear and nonlinear analyses. It details the linear eigenvalue used to model initial local and overall geometric imperfections. The nonlinear material and geometrical analyses related to metal structures are also highlighted in Chapter 4. In addition, the chapter also gives a detailed explanation for the RIKS method used in ABAQUS [1.27] that can accurately model the collapse behavior of metal structural elements.

Chapters 5—7 give illustrative examples for finite element models developed to understand the structural behavior of metal columns, beams, and tubular connections, respectively. These chapters start by a brief introduction to the contents as well as a detailed review on previous investigations on the subject. The chapters also detail the developed finite element models and the results obtained. The presented examples show the effectiveness of finite element models in providing detailed data that complement experimental data in the field. The results are discussed to show the significance of the finite element models in predicting the structural response of different metal structural elements.

Finally, Chapter 8 presents design examples for metal tubular connections. The chapter starts by a brief introduction to the contents. The chapter also details the finite element models developed for the presented metal tubular connections. The design rules specified in current codes of practice for the presented connections are also discussed and detailed in this chapter. At the end of the chapter, comparisons between design predictions and finite element results are presented.

1.2. TYPES OF METAL STRUCTURES

The main objective of this book is to provide a complete piece of work regarding finite element analysis of metal structures. Hence, it is decided to highlight finite element modeling of main metal structural elements, which are columns, beams, and tubular connections. The metal structures cover structures that may be constructed from any metal such as carbon steel, cold-formed steel, stainless steel, aluminum, or any other metals. The aforementioned materials have different stress—strain curves, yield, and post-yield criteria. Figure 1.1 shows examples of stress—strain curves

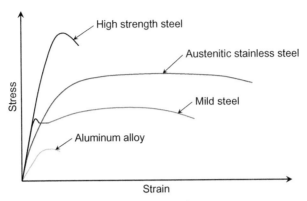

Figure 1.1 Stress–strain curves of different metals.

for some of the aforementioned metals. For example, the stress–strain curves of stainless steel, high strength steel, and aluminum have a rounded behavior with no yield plateau compared with the stress–strain curves of carbon steel as shown in Figure 1.1. Hence, the structural performance of these metal columns, beams, and tubular connections will be different from that of carbon steel. This book provides a detailed description on finite element analysis of columns, beams, and tubular connections that are composed of any metallic materials. It should also be noted that the structural performance of different metals varies at ambient temperature as well as at elevated temperatures. However, this book only focuses on analyzing metal structures at ambient temperature. Furthermore, the finite element analysis of metal structures depends on the type of applied loads. For example, the structural performance of metal structural elements subjected to static loads differs from that subjected to seismic, cyclic, dynamic loads or any other types of loads. However, this book details the finite element analysis of metal structures subjected to static loads or any other loads that can be replaced by equivalent static loads.

Looking at the metal columns analyzed using the finite element method in this book, the columns can be individual metal columns, which represent the cases of metal column test specimens. On the other hand, the columns investigated can be parts of structural metal frames or trusses. The columns presented in this book can have different end boundary conditions that vary from free to fixed-ended columns, different lengths, and different cross sections constructed from hot-rolled, cold-formed, or welded built-up sections. Figure 1.2 shows examples of different column cross sections that can be investigated using finite element

Figure 1.2 Cross sections of some metal columns covered in this book.

analysis covered in this book. The examples of cross sections are square, rectangular, circular, I-shaped, solid, hollow, stiffened, and unstiffened sections.

The metal beams presented in this book using the finite element method can also form single metal beams such as metal beam test specimens. Alternatively, the beams can be part of floor beams used in structural metal frames or framed trusses. Therefore, the beams investigated also can have different end boundary conditions that vary from free to fixed support with or without rigid and semi-rigid internal and end supports. The beams investigated can have different lengths and different

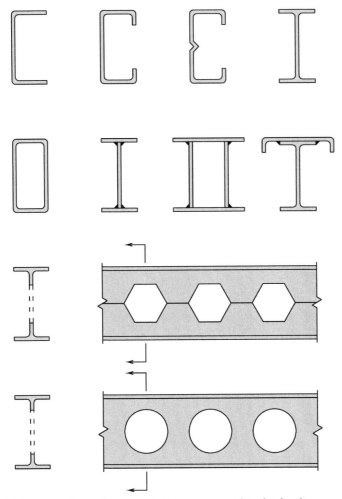

Figure 1.3 Cross sections of some metal beams covered in this book.

cross sections constructed from hot-rolled, cold-formed, or welded built-up sections. Figure 1.3 shows examples of different beam cross sections that can be investigated using finite element analysis. The examples of cross sections include I-shaped, channel, hollow, castellated, cellular, stiffened, and unstiffened sections, as shown in Figure 1.3.

Investigating the interaction between metal columns and beams using finite element analysis is also covered in this book. The beams and columns are the main supporting elements of any metal frames and trusses. By highlighting the structural performance of metal tubular connections,

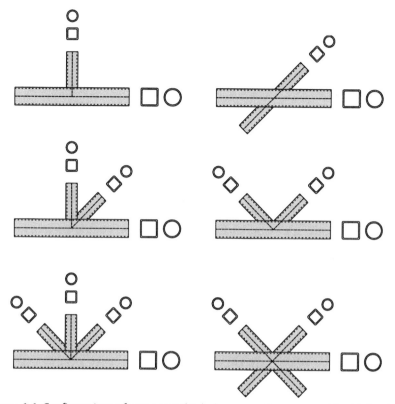

Figure 1.4 Configurations of some metal tubular connections covered in this book.

the building structural behavior can be investigated. The connections investigated can have different boundary conditions at the ends and can be rigid or semi-rigid connections. The tubular connections can have different cross sections constructed from hot-rolled, cold-formed, and welded sections. Figure 1.4 shows examples of different tubular connections that can be investigated using finite element analysis as detailed in this book. The tubular connections comprise square, rectangular, and circular hollow sections.

1.3. EXPERIMENTAL INVESTIGATIONS AND ITS ROLE FOR FINITE ELEMENT MODELING

Experimental investigation plays a major role in finite element analysis. It is important to verify and validate the accuracy of finite element models using test data, particularly nonlinear finite element models. In order to investigate the performance of a structural member, the member must be either tested in laboratory to observe the actual behavior or theoretically analyzed to obtain an exact closed-form solution. Getting an exact solution sometimes becomes very complicated and even impossible in some cases that involve highly nonlinear material and geometry analyses. However, experimental investigations are also costly and time consuming, which require specialized laboratory and expensive equipment as well as highly trained and skilled technician. Without the aforementioned requirements, the test data and results will not be accurate and will be misleading to finite element development. Therefore, accurate finite element models should be validated and calibrated against accurate test results.

Experimental investigations conducted on metal structures can be classified into full-scale and small-scale tests. In structural member tests, full-scale tests are conducted on members that have the same dimensions, material properties, and boundary conditions as that in actual buildings or constructions. On the other hand, small-scale tests are conducted on structural members that have dimensions less proportional to actual dimensions. The full-scale tests are more accurate without the size effect and provide more accurate data compared with small-scale tests; however, they are more expensive in general. Most of the experimental investigations carried out on metal structures are destructive tests in nature. This is attributed to the tests that are carried out until failure or collapse of the member in order to predict the capacity, failure mode, and overall structural member behavior. Tests must be very well planned and sufficiently instrumented to obtain required information. Efficient testing programs must investigate most of the parameters that affect the structural performance of tested specimens. The programs should also include some repeated tests to check the accuracy of the testing procedures. Experimentalists can efficiently plan the required number of tests, position, type, and number of instrumentations as well as significant parameters to be investigated.

Experimental investigations on metal structures are conducted to obtain required information from the tests using proper instrumentations and measurement devices. Although the explanation of various instrumentations and devices is not in the scope of this book, the required information for finite element analysis is highlighted herein. The required information can be classified into three main categories: initial data, material data, and data at the time of experiment. The initial data are obtained from test specimens prior to testing, such as the initial local and overall geometric imperfections, residual stresses, and dimensions of test speci-mens. The material data are conducted on tensile or compression coupon test specimens taken from untested specimens or material tests conducted on whole untested specimens, such as stub column tests, to determine the stress—strain curves of the materials. Knowing the stress—strain curves of materials provide the data regarding the yield stress, ultimate stress, strain at yield, strain at failure, ductility as well as initial modulus of elasticity. Finally, the data at the time of experiment provide the strength of structural test specimens, load—displacement relationships, load—strain relationships, and failure modes. The aforementioned data are examples of the main and commonly needed data for finite element analysis; however, the authors of this book recommend that each experimental investigation should be treated as an individual case and the data required have to be carefully studied to cover all parameters related to the tested structural element.

The tests conducted by Young and Lui [1.28,1.29] on cold-formed high strength stainless steel square and rectangular hollow section columns provided useful and required initial data, material data, and data at the time of experiment for development of finite element model. First, the tests have provided detailed data regarding initial local and overall geometric imperfections as well as residual stresses in the specimens, which represent "initial data." Second, the tests have provided detailed material properties for flat and corner portions of the sections, which represent "material data." Finally, the tests have provided detailed data on the compression column tests, which represent "test data at the time of experiment." Figure 1.5 shows the measured membrane residual stress distributions in cold-formed high strength stainless steel rectangular hollow section. The values of the residual stresses that are "material data" can be incorporated in the finite element model.

1.4. FINITE ELEMENT MODELING OF METAL STRUCTURES

Although extensive experimental investigations were presented in the literature on metal structures, the number of tests on some research topics is still limited. For example, up-to-date, the presented tests (Section 1.3) on cold-formed high strength stainless steel columns carried out by Young and Lui [1.28,1.29] remain pioneer in the field, and there is a lack of test data that highlight different parameters outside the scope of the presented experimental program [1.28,1.29]. The number of tests conducted on a

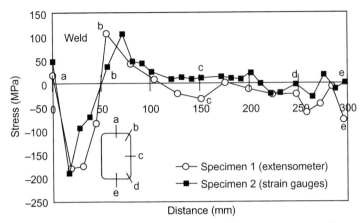

Figure 1.5 Measured membrane residual stress distributions in cold-formed high strength stainless steel tubular section [1.29].

specific research topic in the field of metal structures is limited by many factors. The factors comprise time, costs, labor, capacity of testing frame, capacity of loading jack, measurement equipment, and testing devices. Therefore, numerical investigations using finite element analysis were performed and found in the literature to compensate the lack of test data in the field of metal structures. However, detailed explanation on how successful finite element analysis can provide a good insight into the structural performance of metal structures was not fully addressed as a complete piece of work, which is credited to this book.

Following experimental investigations on metal structures, finite element analyses can be performed and verified against available test results. Successful finite element models are those that are validated against sufficient number of tests, preferably from different sources. Finite element modeling can be extended, once validated, to conduct parametric studies investigating the effects of the different parameters on the behavior and strength of metal structures. The analyses performed in the parametric studies must be well planned to predict the performance of the investigated structural elements outside the ranges covered in the experimental program. The parametric studies will generate more data that fill in the gaps of the test results. Hence, one of the advantages of the finite element modeling is to extrapolate the test data. However, the more significant advantage of finite element modeling is to clarify and explain the test data, which is credited to successful finite element models only. Successful finite element models can critically analyze test results and explain reasons behind failure of metal structures. The successful finite element models can go deeply in the test results to provide deformations, stresses, and strains at different locations in the test specimens, which is very difficult to be determined by instrumentation. The successful finite element models can save future tests in the studied research topic owing to that they can investigate different lengths, boundary conditions, cross sections, geometries, material strengths, and different loading.

As an example on how finite element analysis can generate more data to complement test results, the column tests conducted by Young and Lui [1.28,1.29] were modeled by Ellobody and Young [1.30]. The tested specimens were 15 square and rectangular hollow sections of cold-formed high strength stainless steel columns. The measured initial local and overall geometric imperfections and material nonlinearity of the flat and corner portions of the high strength stainless steel sections were carefully incorporated in the finite element model [1.30]. The column strengths and failure modes as well as the load-shortening curves of the columns were obtained using the finite element model. The validated finite element model [1.30] was used to perform parametric studies involving 42 new columns. The new columns investigated the effects of cross section geometries on the strength and behavior of cold-formed high strength stainless steel columns.

1.5. CURRENT DESIGN CODES

Design guides and specifications are proposed in different countries to define standards of metal structural sections, classification of sections, methods of analysis for structural members under different loading and boundary conditions, design procedures, material strengths, and factors of safety for designers and practitioners. The design guides are commonly based on experimental investigations. Many design formulas specified in current codes of practice are in the form of empirical equations proposed by experts in the field of metal structures. However, the empirical equations only provide guidance for design of metal structural elements in the ranges covered by the specifications. The ranges covered by the specification depend on the number of tests conducted on the metal structural elements at the time of proposing the codes. Since there are continuing progress in research to discover new materials, sections, connections, and different loading, the codes of practice need to update from time to time. Furthermore, test programs on metal structural elements are dependent on limits of the test specimens, loading, boundary conditions, and so on. Therefore, the design equations specified in current codes of practice always have limitations. Finite element analysis can provide a good insight into the behavior of metal structural elements outside the ranges covered by specifications. In addition, finite element analysis can check the validity of the empirical equations for sections affected by nonlinear material and geometry, which may be ignored in the specifications. Furthermore, design guides specified in current codes of practice contain some assumptions based on previous measurements, e.g., assuming values for initial local and overall imperfections in metal structural elements. Also, finite element modeling can investigate the validity of these assumptions. This book addresses the efficiency of finite element analyses, and the numerical results are able to improve design equations in the current codes of practice more accurately. However, it should be noted that there are many specifications developed all over the world for metal structures, such as steel structures, stainless steel structures, cold-formed steel structures, and aluminum structures. It is not the intension to include all these codes of practice in this book. Once again, this book focuses on finite element analysis. Therefore, the book only highlights the codes of practice related to the metal structures that performed finite element analysis.

As an example, the cold-formed high strength stainless steel columns tested by Young and Lui [1.28,1.29] and modeled by Ellobody and Young [1.30] as discussed in Sections 1.3 and 1.4, respectively, were assessed against the predications by the design codes of practice related to cold-formed stainless steel structures. The column test results [1.28,1.29] and finite element analysis results [1.30] were compared with design strengths calculated using the American [1.31], Australian/New Zealand [1.32], and European [1.33] specifications for cold-formed stainless steel structures. Based on the comparison between finite element analysis strengths and design strengths, it was concluded [1.30] that the design rules specified in the American, Australian/New Zealand, and European specifications are generally conservative for cold-formed high strength stainless steel square and rectangular hollow section columns, but unconservative for some of the short columns. It should be noted that this is an example on stainless steel columns only. The finite element analysis can be used to other metal structures. Subsequently, more numerical data can be generated and design equations in current codes of practice can be improved to cope with the advances in technology, materials, and constructions. Due to the advances in technology and materials, new construction materials and new structural sections are being produced. For example, a relatively new type of stainless steel called lean duplex, high strength structural steel having yield stress of 960 MPa or above, and section shape of oval and other shapes are used in construction.

REFERENCES

[1.1] Oden, J. T. Finite element of nonlinear continua. New York: McGraw-Hill, 1972.
[1.2] Kardestuncer, H. Elementary matrix analysis of structures. New York: McGraw-Hill, 1974.
[1.3] Whiteman, J. R. A bibliography for finite elements. London: Academic Press, 1975.
[1.4] Norrie, D. and deVries, G. Finite element bibliography. New York: IFI/Plenum, 1976.
[1.5] Zienkiewicz, O. C. The finite element method. 3rd ed., London: McGraw-Hill, 1977.
[1.6] Cook, R. D. Concepts and applications of finite element analysis. 2nd ed., New York: John Wiley & Sons, 1981.
[1.7] Logan, H. A first course in the finite element method. Boston: PWS Engineering, 1986.
[1.8] Timoshenko, S. P. and Gere, J. M. Theory of elastic stability. New York: McGraw-Hill, 1961.
[1.9] Bresler, B. and Lin, T. Y. Design of steel structures. New York: John Wiley & Sons, 1964.
[1.10] Case, J. and Chiler, A. H. Strength of materials. London: Edward Arnold, 1964.
[1.11] Richards, K. G. Fatigue strength of welded structures. Cambridge: The Welding Institute, 1969.
[1.12] Boyd, G. M. Brittle fracture in steel structures. London: Butterworths, 1970.
[1.13] Steed designer manual. Crosby, Lockwood, London, 1972.
[1.14] Johnson, B. G. Guide to stability design criteria for metal structures. 3rd Ed., New York: John Wiley & Sons, 1976.
[1.15] Trahair, N. S. The behaviour and design of steel structures. London: Chapman and Hall, 1977.
[1.16] Ghali, A. and Neville, A. M. Structural analysis. London: Chapman and Hall, 1978.
[1.17] Pratt, J. L. Introduction of the welding of structural steelwork. London: Constrado, 1979.
[1.18] Horne, M. R. Plastic theory of structures. Oxford: Pergamon Press, 1979.
[1.19] Pask, J. W. Manual on connections for beam and column construction. London: BCSA, 1981.
[1.20] Jenkins, V. M. Structural mechanics and analysis. New York: Van Nostrand-Reinhold, 1982.
[1.21] Holmes, M. and Martin, L. H. Analysis and design of structural connections—reinforced concrete and steel. London: Ellis Horwood, 1983.
[1.22] Bates, W. Design of structural steelwork: Lattice framed industrial buildings. London: Constrado, 1983.
[1.23] Steelwork design guide to BS 5950: Part I, Vol. 1, Section properties, member capacities. The Steel Construction Institute, Ascot, 1987.
[1.24] Steelwork design guide to BS 5950: Part I, Vol. 2, Worked examples. The Steel Construction Institute, Ascot, 1986.
[1.25] MacGinley, T. J. and Ang, T. C. Structural steelwork: Design to limit state theory. Oxford: Butterworth-Heinemann Ltd., 1992.
[1.26] Yu, W. W. Cold-formed steel design. 3rd Edition, Inc., New York: John Wiley & Sons, 2000.

[1.27] ABAQUS Standard User's Manual. Hibbitt, Karlsson and Sorensen, Inc. Vol. 1, 2 and 3, Version 6.8-1, USA, 2008.

[1.28] Young, B. and Lui, W. M. Tests of cold-formed high strength stainless steel compression members. Thin-walled Structures, Elsevier Science, 44(2), 224−234, 2006.

[1.29] Young, B. and Lui, W. M. Behavior of cold-formed high strength stainless steel sections. Journal of Structural Engineering, ASCE, 131(11), 1738−1745, 2005.

[1.30] Ellobody, E. and Young, B. Structural performance of cold-formed high strength stainless steel columns. Journal of Constructional Steel Research, 61(12), 1631−1649, 2005.

[1.31] ASCE. Specification for the design of cold-formed stainless steel structural members. Reston (VA): American Society of Civil Engineers, SEI/ASCE-02, 2002.

[1.32] AS/NZS, Cold-formed stainless steel structures. Australian/New Zealand Standard, Sydney (Australia): Standards Australia, AS/NZS 4673:2001, 2001.

[1.33] EC3. Eurocode 3: design of steel structures—Part 1.4: General rules—Supplementary rules for stainless steels. Brussels: European Committee for Standardization, ENV 1993-4, CEN, 1996.

CHAPTER 2

Review of the General Steps of Finite Element Analysis

2.1. GENERAL REMARKS

This chapter presents a brief review of the finite element method for application on metal structures. The development of the finite element method in the field of structural engineering was credited to the numerical investigations performed by Hernnikoff [2.1] and McHenery [2.2]. The investigations [2.1,2.2] were limited to the use of *one-dimensional* (1D) elements for the evaluation of stresses in continuous structural beams. The investigations [2.1,2.2] were followed by the use of shape functions as a method to obtain approximate numerical investigations as detailed in Ref. [2.3]. Following the study [2.3], the flexibility or force method was proposed [2.4,2.5] mainly for analyzing aircraft structures. *Two-dimensional* (2D) elements were first introduced in Ref. [2.6], where stiffness matrices were derived for truss, beam, and 2D triangular and rectangular elements in plane stress conditions. The study [2.6] has outlined the fundamentals of the stiffness method for predicting the structure stiffness matrix. The development of the finite element method was first introduced by Clough [2.7] where triangular and rectangular elements were used for the analysis of structures under plane stress conditions.

In 1961, Melosh [2.8] developed the stiffness matrix for flat rectangular plate bending elements that was used for the analysis of plate structures. This was followed by developing the stiffness matrix of curved shell elements for the analysis of shell structures as detailed by Grafton and Strome [2.9]. Numerous investigations were developed later on to highlight the finite element analysis of *three-dimensional* (3D) structures as presented in Refs [2.10–2.15]. Most of the aforementioned investigations addressed structures under small strains and small displacements, elastic material, and static loading. Structures that underwent large deflection and buckling analyses were detailed in Refs [2.16,2.17], respectively. Improved numerical techniques for the solution of finite element equations were first addressed by Belytschko [2.18,2.19]. Recent developments in computers have resulted in the finite element method being used to describe complicated structures associated with large number of equations. Numerous *special-purpose* and *general-purpose* programs have been written to analyze various complicated structures with the advent of computers and computational programs. However, to successfully use computers in finite element analyses, it is important to understand the fundamentals of developing finite element models comprising the definition of nodal coordinates, finite elements and how they are connected, material properties of the elements, applied loads, boundary conditions, and the kind of analysis to be performed.

There are two main approaches associated with finite element analyses, which are dependent on the type of results predicted from the analyses. The first approach is commonly known as *force* or *flexibility* method, which considers internal forces are the unknowns of the analyses. On the other hand, the second approach is called the *displacement* or *stiffness* method, which deals with nodal displacement as the unknowns of the analyses. The finite element formulations associated with the two approaches have different matrices that are related to flexibility or stiffness. Previous investigations by Kardestuncer [2.20] have shown that the displacement method is more desirable from computational purposes, and its finite element formulations, for most of the structural analyses, are simple compared with the force method. The majority of general-purpose finite element programs have adopted the displacement method. Therefore, the displacement method will be explained in this chapter. It should be noted that, in the displacement method, the unknown displacements calculated during the analysis are normally called *degrees of freedom*. The degrees of freedom are translational and rotational displacements at each node. The number of degrees of freedom depends on the element type, and hence it is variable from a structure to another.

The finite element method is based on modeling the structure using small interconnected elements called *finite elements* with defined points forming the element boundaries called *nodes*. There are numerous finite elements analyzed in the literature such as bar, beam, frame, solid, and shell elements. The use of any element depends on the type of the structure, geometry, type of analysis, applied loads and boundary conditions, computational time, and data required from the analysis. Each element has its own *displacement function* that describes the displacement within the element in terms of nodal displacement. Every interconnected element has to be linked to other elements simulating the structure directly by sharing the exact boundaries or indirectly through the use of interface nodes, lines, or elements that connect the element with the other elements. The *element stiffness matrices* and finite element equations can be generated by making use of the commonly known stress—strain relationships and direct equilibrium equations. By solving the finite element equations, the unknown displacements can be determined and used to predict different straining actions such as internal forces and bending moments.

The main objective of this chapter is to provide a general review of the main steps of finite element analysis of structures, specifically for metal structures. This chapter introduces the background of the finite element method that was used to write most of the special and general-purpose programs available in the literature. In addition, this chapter reviews different finite element types used to analyze metal structures. The selection of displacement functions, definition of strain—displacement, stress—strain relationships, the formation of element stiffness matrices and equations, the assemblage of these equations, and how the assembled equations are solved for unknowns are also briefed in this chapter. A simplified illustrative example is also included in this chapter to show how these steps are implemented. It is intended not to complicate the derived finite element equations presented in this chapter and not to present more examples since they are previously detailed in numerous finite element books in the literature, with examples given in Refs [1.1—1.7].

2.2. DIVIDING AND SELECTION OF ELEMENT TYPES FOR METAL STRUCTURES

The first step of the finite element method is to divide the structure into small or finite elements defined by nodes located at the element edges. The location of nodes must be chosen to define the positions of changes in the structure. The changes comprise variation of geometry, material, loading, and boundary conditions. The guidelines to divide or mesh different metal structures will be detailed in the coming chapters. It is also important to choose the best finite elements to represent and simulate the structure. *1D elements* or *bar* or *truss elements* shown in Figure 2.1A are often used to model metal trusses. The elements have a cross-sectional area, which is commonly constant, but usually presented by line segments. The simplest line element has two nodes, one at each end, and is called *linear 1D element*. Higher order elements are curved elements that have three or four nodes and are called *quadratic* and *cubic 1D elements*,

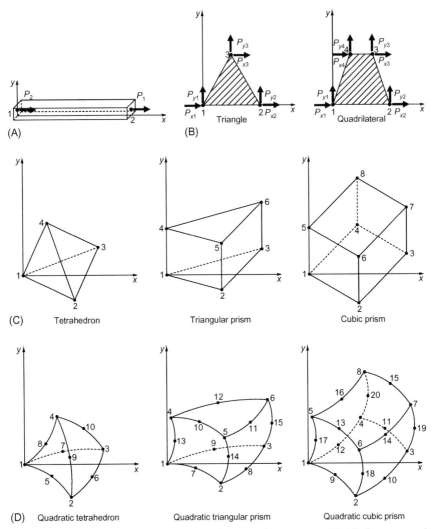

Figure 2.1 Element types commonly used in metal structures: (A) 1D (bar or truss) element, (B) 2D (plane stress or plane strain) elements, (C) linear 3D solid elements, and (D) quadratic 3D solid elements.

respectively. The line element is the simplest form of element and therefore can be used to explain the basic concepts of the finite element method in this chapter. *2D elements* or *plane elements* shown in Figure 2.1B are often used to model metal structures that are loaded by forces in their own plane, commonly named as *plane stress* or *plane strain conditions*. Plane stress elements can be used when the thickness of a metal structure is small relative to its lateral (in-plane) dimensions. The stresses are functions of planar coordinates alone, and the out-of-plane normal and shear stresses are equal to zero. Plane stress elements must be defined in the X−Y plane, and all loading and deformation are also restricted to this plane. This modeling method generally applies to thin, flat bodies. On the other hand, plane strain elements can be used when it can be assumed that the strains in a loaded metal structure are functions of planar coordinates alone and the out-of-plane normal and shear strains are equal to zero. Plane strain elements must be defined in the X−Y plane, and all loading and deformation are also restricted to this plane. This modeling method is generally used for metal structures that are very thick relative to their lateral dimensions. The main 2D elements used are triangular or quadrilateral elements and usually have constant thickness. The simplest 2D elements have nodes at corners and are called *linear 2D elements.* Higher order elements are curved sided elements that have one or two nodes between corners and are called *quadratic* and *cubic 2D elements,* respectively.

Finally, the *3D elements, brick,* or *solid elements,* shown in Figure 2.1C and D, are often used to model metal structures that are loaded by forces in 3D named *three-dimensional stress analysis.* The main elements used are tetrahedral and hexahedral elements and usually used to represent metal structures that have complicated 3D geometry. The simplest 3D elements have nodes at corners as shown in Figure 2.1C and are called *linear 3D elements.* Higher order elements are curved surface elements that have one or two nodes between corners as shown in Figure 2.1D and are called *quadratic* and *cubic 3D elements,* respectively. The *axisymmetric elements* are 3D elements as shown in Figure 2.2 that are formed by rotating a triangle (Figure 2.2A) or quadrilateral (Figure 2.2B) about a fixed axis throughout 360°. They are used to model metal structures that have axisymmetric geometry. The axisymmetric elements are commonly given coordinates using the $r − \theta − z$ domain, where r is the radius from origin to node, θ is the angle from horizontal axis, and z is the vertical coordinate, as shown in Figure 2.2.

From the structural point of view, the element types can be classified mainly to truss, beam, frame, and shell elements. Truss or membrane elements are elements that transmit in-plane forces only (no bending moments) and have no bending stiffness, as shown in Figure 2.3A. The elements are mainly long, slender structural members such as link members and are presented in 1D. Beam elements are the elements that

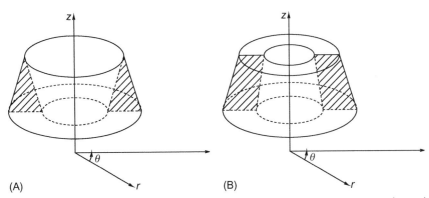

Figure 2.2 Axisymmetric solid elements: (A) develop by rotating a triangle and (B) develop by rotating a quadrilateral.

transfer lateral forces and bending moments. Hence, the deformations associated with beam elements are transverse displacement and rotation, as shown in Figure 2.3B. The dimensions of the cross section are small compared to the dimensions along the axis of the beam. The axial dimension must be interpreted as a global dimension (not the element length), such as distance between supports or distance between gross changes in cross section. The main advantage of beam elements is that they are geometrically simple and have few degrees of freedom. This simplicity is achieved by assuming that the member's deformation can be estimated entirely from variables that are functions of position along the beam axis only. Frame elements are elements that provide efficient modeling for design calculations of frame-like structures composed of initially straight, slender members. They operate directly in terms of axial force, transverse force, and bending moments at the element's end nodes. Hence, the deformations associated with frame elements are axial and transverse displacements and rotation, as shown in Figure 2.3C. Frame elements are two-node, initially straight, slender beam elements intended for use in the analysis of frame-like structures. Similar to beam elements, frame elements are commonly presented in 2D. However, some of the general-purpose programs have the ability to analyze beams and frames in 3D by including additional degrees of freedom for the elements in the plane perpendicular to their plane.

It should be noted that the fundamental assumption used with beam and frame elements' section is that it cannot deform in its own plane. The implications of this assumption should be considered carefully in any use

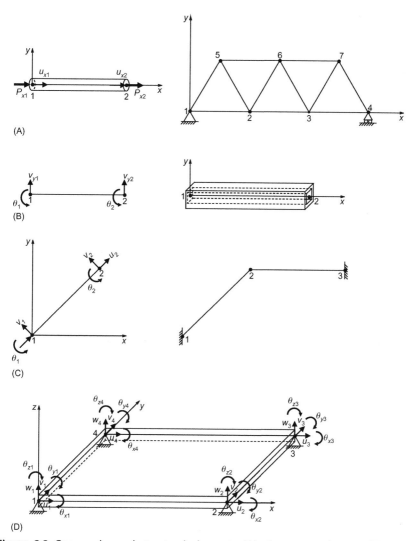

Figure 2.3 Commonly used structural elements: (A) plane truss element, (B) plane beam element, (C) plane frame element, and (D) shell element.

of beam and frame elements, especially for cases involving large amounts of bending or axial tension/compression of non-solid cross sections such as pipes, I-beams, and U-beams. Beam and frame elements' section collapse may occur and result in very weak behavior that is not predicted by the assumptions of beam theory. Similarly, thin-walled, curved pipes exhibit much softer bending behavior than would be predicted by beam theory because the pipe wall readily bends in its own section. This effect, which must generally be considered when designing piping elbows, can be modeled by using shell elements. Shell elements are used to model structures in which one dimension (the thickness) is significantly smaller than the other dimensions. Shell elements use this condition to discretize a structure by defining the geometry at a reference surface, commonly the mid-plane surface. Shell elements have displacement and rotational degrees of freedom at nodes, as shown in Figure 2.3D. The shell section behavior may require numerical integration over the section, which can be linear or nonlinear and can be homogeneous or composed of layers of different material.

The number of elements required to simulate a structure is very important. The more elements used to simulate a structure, the more usable results we get and the more efficiency we obtain to represent the structural behavior. However, the more elements used, the more computational time to perform the finite element analysis. It should be noted that the increase in the number of elements does not increase the accuracy of the results obtained after a certain number, and any increase in the number of the elements would give approximately the same result as shown in Figure 2.4. Hence, the element size (number of elements) has to be carefully decided to give accurate results compared with tests or exact closed-form solution and at the same time to take reasonable computational time.

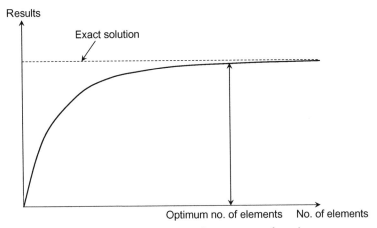

Figure 2.4 Effect of number of elements on the accuracy of results.

2.3. SELECTION OF A DISPLACEMENT FUNCTION

After dividing the structure into suitable finite elements, a *displacement function* or a *shape function* within each element has to be chosen. As mentioned previously, each element type has a certain function that is characteristic to this element. For example, a displacement function for 1D elements is not suitable to represent 2D or 3D elements. The displacement function is defined as the function that describes the displacement within the element in terms of the nodal values of the element. The functions that can be used as shape functions are polynomial functions and may be linear, quadratic, or cubic polynomials. However, trigonometric series can be also used as shape functions. For example, the displacement function of a 2D element is a function of the coordinates in its plane ($X-Y$ plane). The functions are expressed in terms of the nodal unknowns (an x and a y component). The same displacement function can be used to describe the displacement behavior within each of the remaining finite elements of the structure. Hence, the finite element method treats the displacement throughout the whole structure approximately as a discrete model composed of a set of piecewise continuous functions defined within each finite element of the structure.

2.4. DEFINITION OF THE STRAIN–DISPLACEMENT AND STRESS–STRAIN RELATIONSHIPS

The next step, following the selection of a displacement function, is to define the strain–displacement and stress–strain relationships. The relationships depend on the element type and are used to derive the governing equations of each finite element. As an example, a 1D finite element has only one deformation along the axis of the element (x-direction in Figure 2.1A. Assuming that the axial displacement is u, then the axial strain associated with this deformation ε_x can be evaluated as follows:

$$\varepsilon_x = \frac{du}{dx} \tag{2.1}$$

To evaluate the stresses in the element, the stress–strain relationship or constitutive law has to be used. The relationship is also characteristic to the element type and in this simple 1D finite element, Hooke's law can be applied to govern the stress–strain relationship throughout the element as follows:

$$\sigma_x = E\varepsilon_x \tag{2.2}$$

where σ_x is the stress in direction x, which is related to the strain ε_x, and E is the modulus of elasticity.

2.5. DERIVATION OF THE ELEMENT STIFFNESS MATRIX AND EQUATIONS

The next step, following the definition of strain–displacement and stress–strain relationships, is to derive the element *stiffness matrix* and equations that relate nodal forces to nodal displacements. The element stiffness matrix depends on the element type and it is characteristic to the element. The element matrices are commonly developed using *direct equilibrium method* and *work* or *energy methods*. The direct equilibrium method is the simplest approach to derive the stiffness matrix and element equations. The method is based on applying force equilibrium conditions and force–deformation relationships for each finite element. This method is easy to apply for 1D finite elements and becomes mathematically tedious for higher order elements. Therefore, for 2D and 3D finite elements, the work method is easier to apply. The work method is based on the principle of virtual work as detailed in Ref. [2.21]. Both the direct equilibrium and work methods will yield the same finite element equations relating the nodal forces with nodal displacements as follows:

$$\{f\} = [k]\{d\} \tag{2.3}$$

where $\{f\}$ is the vector of nodal forces, $[k]$ is the finite element stiffness matrix, and $\{d\}$ is the vector of unknown finite element nodal degrees of freedom or displacements. The formulation of the aforementioned finite element matrices will be explained in this chapter by an illustrative example for 1D finite elements. Similar approach can be used for any other elements as detailed in Refs [1.1–1.7].

2.6. ASSEMBLAGE OF ELEMENT EQUATIONS

Following the derivation of the individual element stiffness matrix and equations of each finite element of the structure, the global stiffness matrix and equation of the whole structure can be assembled. The assemblage of the global stiffness matrix and equation is generated by adding and superimposing the individual matrices and equations using the *direct stiffness method*, which is based on the equilibrium of nodal forces. The direct stiffness method is based on the fact that, for any structure in equilibrium, the nodal forces and displacements must be in continuity and compatibility in the individual finite element as well as in the whole structure. The global finite element equation can be expressed in matrix form as follows:

$$\{F\} = [K]\{d\} \tag{2.4}$$

where $\{F\}$ is the assembled vector of the whole structure global nodal forces, $[K]$ is the whole structure assembled global stiffness matrix, and $\{d\}$ is the assembled vector of the whole structure global unknown nodal degrees of freedom or displacements. It should be noted that Eq. (2.4) must be modified to account for the boundary conditions or support constraints. The modification will be explained in the illustrative example detailed in this chapter. Also, it should be noted that Eq. (2.3) may be evaluated for each finite element with respect to local coordinate system. However, the assembled equation (2.4) must be evaluated with respect to a unique generalized coordinate system. Hence, *transformation matrices* must be used to relate local coordinates to general coordinate systems, as detailed in Refs [1.1–1.7].

2.7. SOLVING THE ASSEMBLED EQUATIONS FOR THE UNKNOWNS

Solving Eq. (2.4) will result in the evaluation of the unknown nodal degrees of freedom or generalized displacements. The equation can be solved using algebraic procedures such as elimination or iterative methods detailed in Refs [1.1–1.7]. The calculated unknown nodal degrees of freedom (translational displacements and rotations) can be used to evaluate all required variables in the structure such as stresses, strains, bending moments, shear forces, axial forces, and reactions. The evaluation of the aforementioned variables can be used to design the structure and to define its failure modes and positions of maximum and minimum deformations and stresses.

2.7.1 An Illustrative Example

The metal structure shown in Figure 2.5A is a structural bar (fixed free-edged structure) having linear elastic material properties, two equal length parts $(L_1 = L_2 = L)$ of different cross-sectional areas $(A_1$ and $A_2)$ and different moduli of elasticity (E_1 and E_2). The structural bar is loaded at its free edge with a load P_3. At the first step, the bar is divided into two elements (1−2 and 2−3). The nodes defining the elements (1, 2, and 3) are located at the positions of the change in boundary conditions, geometry and loading, respectively. As a revision, the previously detailed Refs [1.1−1.7] derivation of the finite element equations governing a single finite element 1−2, shown in Figure 2.5B, can be summarized in the following sections. To represent the deformations within the 1D finite element 1−2 in terms of nodal displacement, a linear polynomial shape function can be expressed as in Eq. (2.5):

$$u = a_1 + a_2 x \tag{2.5}$$

where u is the axial deformation in x-direction, a_1 and a_2 are coefficients that are equal to the number of degrees of freedom, 2 for this 1D element. To express the function u in terms of nodal displacements d_{1x} and d_{2x}, we can solve Eq. (2.5) to obtain the equation coefficients as follows:

At $x = 0.0$, $u = d_{1x}$, and by substituting in Eq. (2.5), $a_1 = d_{1x}$.

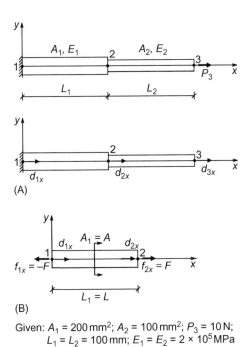

Given: $A_1 = 200\,\text{mm}^2$; $A_2 = 100\,\text{mm}^2$; $P_3 = 10\,\text{N}$;
$L_1 = L_2 = 100\,\text{mm}$; $E_1 = E_2 = 2 \times 10^5\,\text{MPa}$

Figure 2.5 Finite element analysis of a structural bar: (A) structural bar and (B) finite element 1−2.

Also, at $x = L$, $u_2 = d_{2x}$ and once again by substituting in Eq. (2.5), $d_{2x} = d_{1x} + a_2 L$ and $a_2 = (d_{2x} - d_{1x})/L$. By substituting the values of a_1 and a_2 in Eq. (2.5), we can express the deformations within element 1−2 in terms of nodal displacements at nodes 1 and 2 as in Eq. (2.6):

$$u = d_{1x} + (d_{2x} - d_{1x})x/L \tag{2.6}$$

To derive the element stiffness matrix, we have to define the strain−displacement and stress−strain relationships. The axial strain (ε_x) for this 1D element can be expressed as the difference between nodal displacement divided by the element length L as in Eq. (2.7).

$$\varepsilon_x = \frac{du}{dx} = (d_{2x} - d_{1x})/L \tag{2.7}$$

Now, we can use Hooke's law to govern the stress−strain relationship throughout the element as expressed in Eq. (2.8):

$$\sigma_x = E\varepsilon_x \tag{2.8}$$

where, σ_x is the axial stress and E is Young's modulus.

The nodal forces can be now calculated by multiplying the cross-sectional area A by the axial stress σ_x and substituting the axial strain from Eq. (2.7) as in Eq. (2.9):

$$F = A\sigma_x = AE\varepsilon_x = \frac{EA}{L}(d_{2x} - d_{1x}) \tag{2.9}$$

We can now write the nodal force equations by assuming a sign convention that the force at node 1 (f_{1x}) is equal to $-F$ and that at node 2 (f_{2x}) is equal to F. By substituting the nodal forces in Eq. (2.9), we can write the nodal forces f_{1x} and f_{2x} as follows:

$$f_{1x} = \frac{EA}{L}(d_{1x} - d_{2x}) \tag{2.10}$$

$$f_{2x} = \frac{EA}{L}(d_{2x} - d_{1x}) \tag{2.11}$$

Equations (2.10 and 2.11) can be written in matrix form as in Eq. (2.12) and in a compact form as expressed in Eq. (2.13).

$$\begin{Bmatrix} f_{1x} \\ f_{2x} \end{Bmatrix} = \frac{EA}{L} \begin{bmatrix} 1 & -1 \\ -1 & 1 \end{bmatrix} \begin{Bmatrix} d_{1x} \\ d_{2x} \end{Bmatrix} = \begin{bmatrix} k_{11} & k_{12} \\ k_{21} & k_{22} \end{bmatrix} \begin{Bmatrix} d_{1x} \\ d_{2x} \end{Bmatrix} \tag{2.12}$$

$$\{f\} = [k]\{d\} \tag{2.13}$$

We can now easily analyze the structural bar as shown in Figure 2.5A given that the cross-sectional areas A_1 and A_2 are 200 and 100 mm^2, respectively, the moduli of elasticity $E_1 = E_2 = E = 2 \times 10^5$ N/mm^2, the force P_3 is equal to 10 N, and finally the lengths $L_1 = L_2 = L = 100$ mm. The stiffness matrices for finite elements 1–2 and 2–3 ($[k_{12}]$ and $[k_{23}]$) can be evaluated as follows:

$$[k_{12}] = \frac{E_1 A_1}{L_1}\begin{bmatrix} 1 & -1 \\ -1 & 1 \end{bmatrix} = \frac{200,000 \times 200}{100}\begin{bmatrix} 1 & -1 \\ -1 & 1 \end{bmatrix} = 10^5\begin{bmatrix} 4 & -4 \\ -4 & 4 \end{bmatrix} = \begin{bmatrix} k_{11} & k_{12} \\ k_{21} & k_{22} \end{bmatrix}$$

$$[k_{23}] = \frac{E_2 A_2}{L_2}\begin{bmatrix} 1 & -1 \\ -1 & 1 \end{bmatrix} = \frac{200,000 \times 100}{100}\begin{bmatrix} 1 & -1 \\ -1 & 1 \end{bmatrix} = 10^5\begin{bmatrix} 2 & -2 \\ -2 & 2 \end{bmatrix} = \begin{bmatrix} \overline{k}_{22} & \overline{k}_{23} \\ k_{32} & k_{33} \end{bmatrix}$$

The two stiffness matrices $[k_{12}]$ and $[k_{23}]$ can now be assembled to form the global stiffness matrix of the structural bar $[K]$ noting that there are the stiffness k_{13} and k_{31} as follows:

$$[K] = \begin{bmatrix} k_{11} & k_{12} & k_{13} \\ k_{21} & k_{22}+\overline{k}_{22} & k_{23} \\ k_{31} & k_{32} & k_{33} \end{bmatrix} = \begin{bmatrix} k_{11} & k_{12} & 0 \\ k_{21} & k_{22}+\overline{k}_{22} & k_{23} \\ 0 & k_{32} & k_{33} \end{bmatrix} = 10^5 \begin{bmatrix} 4 & -4 & 0 \\ -4 & 6 & -2 \\ 0 & -2 & 2 \end{bmatrix}$$

The global force vector $\{F\}$ and displacement vector $\{d\}$ can now be written as follows:

$$\{F\} = \begin{Bmatrix} f_{1x} \\ f_{2x} \\ f_{3x} \end{Bmatrix} = \begin{Bmatrix} R_1 \\ 0 \\ 10 \end{Bmatrix}$$

$$\{d\} = \begin{Bmatrix} d_{1x} \\ d_{2x} \\ d_{3x} \end{Bmatrix} = \begin{Bmatrix} 0 \\ d_{2x} \\ d_{3x} \end{Bmatrix}$$

The finite element governing equation of the structural bar can now be written and solved for unknown displacements (d_{2x} and d_{3x}) as follows:

$$\begin{Bmatrix} R_1 \\ 0 \\ 10 \end{Bmatrix} = 10^5 \begin{bmatrix} 4 & -4 & 0 \\ -4 & 6 & -2 \\ 0 & -2 & 2 \end{bmatrix} \begin{Bmatrix} 0 \\ d_{2x} \\ d_{3x} \end{Bmatrix}$$

$$\begin{Bmatrix} 0 \\ 10 \end{Bmatrix} = 10^5 \begin{bmatrix} 6 & -2 \\ -2 & 2 \end{bmatrix} \begin{Bmatrix} d_{2x} \\ d_{3x} \end{Bmatrix}$$

$$0 = 10^5(6d_{2x} - 2d_{3x}) \quad \therefore 3d_{2x} = d_{3x}$$

$$10 = 10^5(-2d_{2x} + 2d_{3x}) \quad \therefore 10 = 10^5(-2d_{2x} + 6d_{2x})$$

$$\therefore d_{2x} = 2.5 \times 10^{-5} \text{ mm} \quad \text{and} \quad d_{3x} = 7.5 \times 10^{-5} \text{ mm}$$

We can now obtain all required information regarding the structural bar such as unknown reaction R_1, element strains, and stresses as follows:

$$R_1 = 10^5 \times (-4) \times d_{2x} = 10^5 \times (-4) \times 2.5 \times 10^{-5} = -10 \text{ N}$$

$$\varepsilon_{1x} = \frac{d_{2x} - d_{1x}}{L_1} = \frac{2.5 \times 10^{-5} - 0}{100} = 2.5 \times 10^{-7}$$

$$\varepsilon_{2x} = \frac{d_{3x} - d_{2x}}{L_2} = \frac{7.5 \times 10^{-5} - 2.5 \times 10^{-5}}{100} = 5 \times 10^{-7}$$

$$\sigma_{1x} = E_1\varepsilon_{1x} = 2 \times 10^5 \times 2.5 \times 10^{-7} = 0.05 \text{ N/mm}^2$$

$$\sigma_{2x} = E_2\varepsilon_{2x} = 2 \times 10^5 \times 5 \times 10^{-7} = 0.1 \text{ N/mm}^2$$

The same approach used to analyze 1D truss elements can be used to analyze beam, frame, shell, and solid structural elements as detailed in Refs [1.1–1.7]. The finite element equations and matrices of these elements will become more complicated as the number of degrees of freedom increases. Also, solving the finite element governing equations of higher order elements and obtaining unknown reactions, element strains and element stresses will be more complicated. However, these complicated equations and its solving techniques have been very easy nowadays because of the advances in computers.

REFERENCES

[2.1] Hrennikoff, A. Solution of problems in elasticity by the frame work method. Journal of Applied Mechanics, 8(4), 169−175, 1941.

[2.2] McHenry, D. A lattice analogy for the solution of plane stress problems. Journal of Institution of Civil Engineers, 21, 59−82, 1943.

[2.3] Courant, R. Variational methods for the solution of problems of equilibrium and vibrations. Bulletin of the American Mathematical Society, 49, 1−23, 1943.

[2.4] Levy, S. Computation of influence coefficients for aircraft structures with discontinuities and sweepback. Journal of Aeronautical Sciences, 14(10), 547−560, 1947.

[2.5] Levy, S. Computation of influence coefficients for delta wings. Journal of Aeronautical Sciences, 20(7), 449−454, 1953.

[2.6] Turner, M. G., Clough, R. W., Martin, H. C. and Topp, L. J. Stiffness and deflection analysis of complex structures. Journal of Aeronautical Sciences, 23(9), 805−824, 1956.

[2.7] Clough, R. W. The finite element method in plane stress analysis. Proceedings of American Society of Civil Engineers, 2nd Conference on Electronic computation, Pittsburg, Pa., 345−378, 1960.

[2.8] Melosh, R. J. A stiffness matrix for the analysis of thin plates in bending. Journal of Aerospace Sciences, 28(1), 34−42, 1961.

[2.9] Grafton, P. E. and Strome, D. R. Analysis of axisymmetric shells by the direct stiffness method. Journal of the American Institute of Aeronautics and Astronautics, 1(10), 2342−2347, 1963.

[2.10] Martin, H. C. Plane elasticity problems and the direct stiffness method. The Trend in Engineering, 13, 5−19, 1961.

[2.11] Gallagher, R. H., Padlog, J. and Bijlaard, P. P. Stress analysis of heated complex shapes. Journal of the American Rocket Society, 32, 700−707, 1962.

[2.12] Melosh, R. J. Structural analysis of solids. Journal of the Structural Division, Proceedings of the American Society of Civil Engineers, 205−223, 1963.

[2.13] Argyris, J. H. Recent advances in matrix methods of structural analysis. Progress in Aeronautical Sciences, 4, New York: Pergamon Press, 1964.

[2.14] Clough, R. W. and Rashid, Y. Finite element analysis of axisymmetric solids. Journal of the Engineering Mechanics Division, Proceedings of the American Society of Civil Engineers, 91, 71−85, 1965.

[2.15] Wilson, E. L. Structural analysis of axisymmetric solids. Journal of the American Institute of Aeronautics and Astronautics, 3(12), 2269−2274, 1965.

[2.16] Turner, M. G., Dill, E. H., Martin, H. C. and Melosh, R. J. Large deflections of structures subjected to heating and external loads. Journal of Aeronautical Sciences, 27(2), 97−107, 1963.

[2.17] Gallagher, R. H. and Padlog, J. Discrete element approach to structural stability analysis. Journal of the American Institute of Aeronautics and Astronautics, 1(6), 1437−1439, 1963.

[2.18] Belytschko, T. A survey of numerical methods and computer programs for dynamic structural analysis. Nuclear Engineering and Design, 37(1), 23−34, 1976.

[2.19] Belytschko, T. Efficient large-scale nonlinear transient analysis by finite elements. International Journal of Numerical Methods in Engineering, 10(3), 579−596, 1981.

[2.20] Kardestuncer, H. Elementary matrix analysis of structures. New York: McGraw-Hill, 1974.

[2.21] Oden, J. T. and Ripperger, E. A. Mechanics of elastic structures. New York: McGraw-Hill, 1981.

CHAPTER 3

Finite Element Modeling

3.1. GENERAL REMARKS

The brief revision of the finite element method is presented in Chapter 2. It is now possible to detail the main parameters affecting finite element modeling and simulation of different metal structural members, which is highlighted in this chapter. The chapter provides useful guidelines on how to choose the best finite element type and mesh to represent metal columns, beams and beam columns, and connections. The behavior of different finite elements, briefed in Chapter 2, is analyzed in this chapter to assess their suitability for simulating the structural member. There are many parameters that control the choice of finite element type and mesh such as the geometry, cross section classification, loading, and boundary conditions of the structural member. The aforementioned issues are also covered in this chapter. Accurate finite element modeling depends on the efficiency in simulating the nonlinear material behavior of metal structural members. This chapter shows how to correctly represent different linear and nonlinear regions in the stress—strain curves of metal structures. Most of metal structures have initial local and overall geometric imperfections as well as residual stresses as a result of the manufacturing process. Ignoring the simulation of these initial imperfections and residual stresses would result in poor finite element models that are unable to describe the performance of the metal structure. The correct simulation of different initial geometric imperfections and residual stresses is also addressed in this chapter. In addition, there are different loads and boundary conditions applied to metal columns, beams and beam columns, and connections. Improper simulation of applied loads and boundary conditions on a structural member would not provide an accurate finite element model. Therefore, correct simulation of different loads and boundary conditions that are commonly associated with metal structural members is highlighted in this chapter. Furthermore, the chapter presents examples of finite element models developed in the literature and successfully simulated the performance of different structures. It should be noted that the sections described in this chapter detail the finite element modeling using any software and any finite element package.

Finite Element Analysis and Design of Metal Structures
DOI: http://dx.doi.org/10.1016/B978-0-12-416561-8.00003-2

However, as an example, the use of ABAQUS software [1.27] to simulate different metal structures is detailed in this chapter.

3.2. CHOICE OF ELEMENT TYPE FOR METAL STRUCTURES

To explain how to choose the best finite element type to simulate the behavior of a metal structure, let us start with modeling a stainless steel column having a rectangular hollow section as shown in Figure 3.1. The first step is to look into the classification of the cross section that is normally specified in all current codes of practice. There are three commonly known cross section classifications that are compact, noncompact, and slender sections. Compact sections have a thick plate thickness and can develop their plastic moment resistance without the occurrence of local buckling. Noncompact sections are sections in which the stress in the extreme fibers can reach the yield stress, but local buckling is liable to prevent development of the plastic moment resistance. Finally, slender sections are those sections in which local buckling will occur in one or more parts of the cross section before reaching the yield strength. Compact sections in 3D can be modeled either using solid elements or

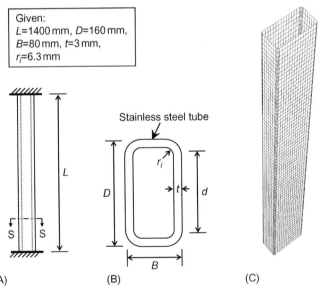

Figure 3.1 Example of a fixed-ended rectangular hollow section column. (A) Fixed-ended rectangular hollow section column. (B) Rectangular hollow section (section S-S). (C) Finite element mesh.

shell elements that are able to model thick sections. However, noncompact and slender sections are only modeled using shell elements that are able to model thin sections. It should be noted that many general-purpose programs have shell elements that are used to simulate thin and thick sections. Finite element models are normally developed to perform different analyses and parametric studies on different cross sections; hence, it is recommended to choose shell elements in modeling the rectangular hollow section column as shown in Figure 3.1.

Let us take a look in detail and classify shell elements commonly used in modeling structural members. There are two main shell element categories known as *conventional* and *continuum shell elements*, examples shown in Figure 3.2. *Conventional shell elements* cover elements used for 3D shell geometries, elements used for axisymmetric geometries, and elements used for stress—displacement analysis. The conventional shell elements can be classified as thick shell elements, thin shell elements, and general-purpose shell elements that can be used for the analysis of thick or thin shells. Conventional shell elements have six degrees of freedom per node; however, it is possible to have shells with five degrees of freedom per node. Numerical integration is normally used to predict the behavior within the shell element. Conventional shell elements can use *full* or

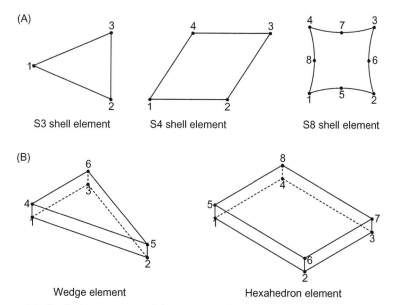

Figure 3.2 Shell element types: (A) conventional shell elements and (B) continuum shell elements.

reduced numerical integration, as shown in Figure 3.3. Reduced integration shell elements use lower order integration to form the element stiffness. However, the mass matrix and distributed loadings are still integrated exactly. Reduced integration usually provides accurate results provided that the elements are not distorted or loaded in in-plane bending. Reduced integration significantly reduces running time, especially in three dimensions. Shell elements are commonly identified based on the number of element nodes and the integration type. Hence, a shell element S8 means a stress—displacement shell having eight nodes with full integration while a shell element S8R means a stress—displacement shell having eight nodes with reduced integration. On the other hand, *continuum shell elements* are general-purpose shells that allow finite membrane deformation and large rotations and, thus, are suitable for nonlinear geometric analysis. These elements include the effects of transverse shear deformation and thickness change. Continuum shell elements employ first-order layer-wise composite theory and estimate through-thickness section forces from the initial elastic moduli. Unlike conventional shells, continuum shell elements can be stacked to provide more refined through-thickness response. Stacking continuum shell elements allows for a richer transverse shear stress and force prediction. It should be noted

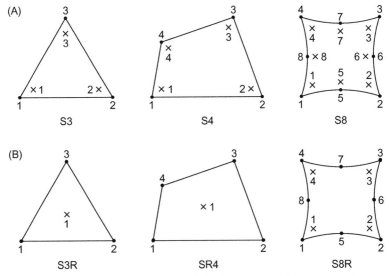

Figure 3.3 Full and reduced integration of shell elements: (A) full integration and (B) reduced integration.

that most metal structures are modeled using conventional shell elements and hence they are detailed in this book.

General-purpose conventional shell elements allow transverse shear deformation. They use thick shell theory as the shell thickness increases and become discrete Kirchhoff thin shell elements as the thickness decreases. The transverse shear deformation becomes very small as the shell thickness decreases. Examples of these elements are S3, S3R, S4, and S4R shells. Thick shells are needed in cases where transverse shear flexibility is important and second-order interpolation is desired. When a shell is made of the same material throughout its thickness, this occurs when the thickness is more than about 1/15 of a characteristic length on the surface of the shell, such as the distance between supports. An example of thick elements is S8R. Thin shells are needed in cases where transverse shear flexibility is negligible and the Kirchhoff constraint must be satisfied accurately (i.e., the shell normal remains orthogonal to the shell reference surface). For homogeneous shells, this occurs when the thickness is less than about 1/15 of a characteristic length on the surface of the shell, such as the distance between supports. However, the thickness may be larger than 1/15 of the element length.

Conventional shell elements can also be classified as *finite-strain* and *small-strain shell elements*. Element types S3, S3R, S4, and S4R account for finite membrane strains and arbitrarily large rotations; therefore, they are suitable for large-strain analysis. On the other hand, small-strain shell elements such as S8R shell elements are used for arbitrarily large rotations but only small strains. The change in thickness with deformation is ignored in these elements. For conventional shell elements used in ABAQUS [1.27], we must specify a section Poisson's ratio as part of the shell section definition to allow for the shell thickness in finite-strain elements to change as a function of the membrane strain. If the section Poisson's ratio is defined as zero, the shell thickness will remain constant and the elements are, therefore, suited for small-strain, large-rotation analysis. The change in thickness is ignored for the small-strain shell elements in ABAQUS [1.27].

Conventional reduced integration shell elements can be also classified based on the number of degrees of freedom per node. Hence, there are two types of conventional reduced integration shell elements known as *five-degrees* and *six-degrees of freedom shells*. Five-degrees of freedom conventional shells have five degrees of freedom per node, which are three translational displacement components and two in-plane rotation components. On the other hand, six-degrees of freedom shells have six degrees of freedom per node, which are three translational displacement components and three rotation components. The number of degrees of freedom per node is commonly denoted in the shell name by adding digit 5 or 6 at the end of the reduced integration shell element name. Therefore, reduced integration shell elements S4R5 and S4R6 have five and six degrees of freedom per node, respectively. The elements that use five degrees of freedom per node such as (S4R5 and S8R5) can be more economical. However, they are suitable only for thin shells and they cannot be used for thick shells. The elements that use five degrees of freedom per node cannot be used for finite-strain applications, although they model large rotations with small strains accurately.

There are a number of issues that must be considered when using shell elements. Both S3 and S3R refer to the same three-node triangular shell element. This element is a degenerated version of S4R that is fully compatible with S4 and S4R elements. S3 and S3R provide accurate results in most loading situations. However, because of their constant bending and membrane strain approximations, high mesh refinement may be required to capture pure bending deformations or solutions to problems involving high strain gradients. Curved elements such as S8R5 shell elements are preferable for modeling bending of a thin curved shell. Element type S8R5 may give inaccurate results for buckling problems of doubly curved shells due to the fact that the internally defined integration point may not be positioned on the actual shell surface. Element type S4 is a fully integrated, general-purpose, finite-membrane-strain shell element. Element type S4 has four integration locations per element compared with one integration location for S4R, which makes the element computation more expensive. S4 is compatible with both S4R and S3R. S4 can be used in areas where greater solution accuracy is required, or for problems where in-plane bending is expected. In all of these situations, S4 will outperform element type S4R.

Based on the previous survey of conventional shell elements, we can find that the general-purpose conventional shell elements S4/S4R can be used effectively to model a metal column having a hollow section. The S3/S3R can also be used in combination with the S4/S4R shells to model curved corners of the hollow section. The elements can be used to model different compact, noncompact, and slender cross sections. Figure 3.1 shows the finite element mesh with shell elements used by Ellobody and Young [1.30] for a fixed-ended rectangular hollow section column having a length (L) of 1400 mm, a depth (D) of 160 mm, a width (B) of 80 mm, plate thickness (t) of 3 mm, and internal corner radius (r_i) of 6.3 mm. The authors have used S4R elements to model the flat and curved portions of the whole column.

As mentioned earlier, metal structures that are composed of compact sections can be modeled using solid elements. Solid or continuum elements are volume elements that do not include structural elements such as beams, shells, and trusses. The elements can be composed of a single homogeneous material or can include several layers of different materials for the analysis of laminated composite solids. The naming conventions for solid elements depend on the element dimensionality, number of nodes in the element, and integration type. For example, C3D8R elements are continuum elements (C), 3D elements having eight nodes with reduced integration (R). Solid elements provide accurate results if not distorted, particularly for quadrilaterals and hexahedra, as shown in Figure 2.1. The triangular and tetrahedral elements are less sensitive to distortion. Solid elements can be used for linear analysis and for complex nonlinear analyses involving stress, plasticity, and large deformations. Solid element library includes first-order (linear) interpolation elements and second-order (quadratic) interpolation elements commonly in three dimensions. Tetrahedral, triangular prisms, and hexahedra (bricks) are very common 3D elements, as shown in Figure 2.1. Modified second-order triangular and tetrahedral elements as well as reduced integration solid elements can be also used. First-order plane strain, axisymmetric quadrilateral, and hexahedral solid elements provide constant volumetric strain throughout the element, whereas second-order elements provide higher accuracy than first-order elements for smooth problems that do not involve severe element distortions. They capture stress concentrations more effectively and are better for modeling geometric features. They can model a curved surface with fewer elements. Finally, second-order elements are very effective in bending-dominated problems. First-order triangular and tetrahedral elements should be avoided as much as possible in stress analysis problems; the elements are overly stiff and exhibit slow convergence with mesh refinement, which is especially a problem with first-order tetrahedral elements. If they are required, an extremely fine mesh may be needed to obtain results with sufficient accuracy.

Similar to the behavior of shells, reduced integration can be used with solid elements to form the element stiffness. The mass matrix and distributed loadings use full integration. Reduced integration reduces running time, especially in 3D. For example, element type C3D20 has 27 integration points, while C3D20R has 8 integration points only. Therefore, element assembly is approximately 3.5 times more costly for C3D20 than for C3D20R. Second-order reduced integration elements generally provide accurate results than the corresponding fully integrated elements. However, for first-order elements, the accuracy achieved with full versus reduced integration is largely dependent on the nature of the problem. Triangular and tetrahedral elements are geometrically flexible and can be used in many models. It is very convenient to mesh a complex shape with triangular or tetrahedral elements. A good mesh of hexahedral elements usually provides a solution with equivalent accuracy at less cost. Quadrilateral and hexahedral elements have a better convergence rate than triangular and tetrahedral elements. However, triangular and tetrahedral elements are less sensitive to initial element shape, whereas first-order quadrilateral and hexahedral elements perform better if their shape is approximately rectangular. First-order triangular and tetrahedral elements are usually overly stiff, and fine meshes are required to obtain accurate results. For stress—displacement analyses, the first-order tetrahedral element C3D4 is a constant stress tetrahedron, which should be avoided as much as possible. The element exhibits slow convergence with mesh refinement. This element provides accurate results only in general cases with very fine meshing. Therefore, C3D4 is recommended only for filling in regions of low stress gradient to replace the C3D8 or C3D8R elements, when the geometry precludes the use of C3D8 or C3D8R elements throughout the model. For tetrahedral element meshes, the second-order or the modified tetrahedral elements such as C3D10 should be used. Similarly, the linear version of the wedge element C3D6 should generally be used only when necessary to complete a mesh, and, even then, the element should be far from any area where accurate results are needed. This element provides accurate results only with very fine meshing. A solid section definition is used to define the section properties of solid elements. A material definition must be defined with the solid section definition, which is assigned to a region in the finite element model.

As mentioned previously in Chapter 2, plane-stress and plane-strain structures can be modeled using 2D solid elements. The naming conventions for the elements depend on the element type (PE or PS) for (plane strain or plane stress), respectively, and number of nodes in the element. For example, CPE3 elements are continuum (C), plane strain (PE) linear elements having three nodes, as shown in Figure 2.1. The elements have two active degrees of freedom per node in the element plane. Quadratic 2D elements are suitable for curved geometry of structures. Structural metallic link members and metallic truss members can be modeled using 1D solid elements. The naming conventions for 1D solid elements depend on the number of nodes in the element. For example C1D3 elements are continuum (C) elements having three nodes. The elements have one active degree of freedom per node.

Axisymmetric solid elements are 3D elements that are used to model metal structures that have axisymmetric geometry. The element nodes are commonly using cylindrical coordinates (r, θ, z), where r is the radius from origin (coordinate 1), θ is the angle in degrees measured from horizontal axis (coordinate 2), and z is the perpendicular dimension (coordinate 3) as shown in Figure 3.4. Coordinate 1 must be greater than or equal to zero. Degree of freedom 1 is the translational displacement along the radius (u_r), and degree of freedom 2 is the translational displacement along the perpendicular direction (u_z). The naming conventions for axisymmetric solid elements with nonlinear asymmetric deformation depend on the number

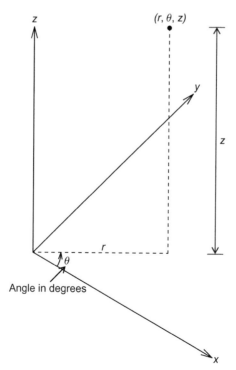

Figure 3.4 Cylindrical coordinates for axisymmetric solid elements.

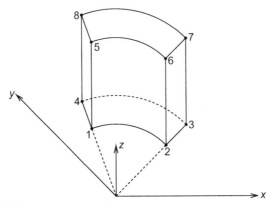

Figure 3.5 CAXA8R axisymmetric solid elements.

of nodes in the element and integration type. For example, CAXA8R elements are continuum (C) elements, axisymmetric solid elements with nonlinear asymmetric deformation (AXA) having eight nodes with reduced (R) integration as shown in Figure 3.5. Stress—displacement axisymmetric solid elements without twist have two active degrees of freedom per node.

3.3. CHOICE OF FINITE ELEMENT MESH FOR METAL STRUCTURES

After choosing the best finite element type to model a metal structural member, we need to look into the geometry of the metal structural member to decide the best finite element mesh. Normally, most cold-formed and hot-rolled metal structural members have flat and curved regions. Therefore, the finite element mesh has to cover both flat and curved regions. Also, most metal structural members have short dimensions, which are commonly the lateral dimensions of the cross section, and long dimensions, which are the longitudinal axial dimension of the structural member that defines the structural member length. Therefore, the finite element mesh has to cover both lateral and longitudinal regions of the structural member. Once again, let us mesh the cold-formed stainless steel hollow section column shown in Figure 3.1. We have already explained that general-purpose conventional quadrilateral shell elements S4R can be used to model the hollow section column effectively as used by Ellobody and Young [1.30]. To mesh the column (Figure 3.1) correctly, we have to start with a short dimension for the chosen shell element and decide the best *aspect ratio*. The aspect ratio is defined as the

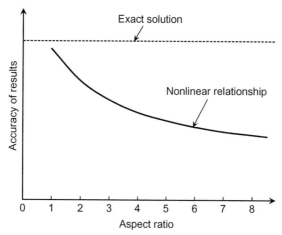

Figure 3.6 Effect of aspect ratio of finite elements on the accuracy of results.

ratio of the longest dimension to the shortest dimension of a quadrilateral finite element. As the aspect ratio is increased, the accuracy of the results is decreased. The aspect ratio should be kept approximately constant for all finite element analyses performed on the column. Therefore, most general-purpose finite element software specify a maximum value for the aspect ratio that should not be exceeded; otherwise, the results will be inaccurate. Figure 3.6 presents a schematic diagram showing the effect of aspect ratio on the accuracy of results. The best aspect ratio is 1, and the maximum value, as an example the value recommended by ABAQUS [1.27], is 5. It should be noted that the smaller the aspect ratio, the larger the number of elements and the longer the computational time. Hence, it is recommended to start with an aspect ratio of 1 and mesh the whole column and compare the numerical results against test results or exact closed-form solutions. Then we can repeat the procedure using aspect ratios of 2 and 3 and plot the three numerical results against test results or exact solutions. After that, we can go back and choose different short dimensions smaller or larger than that initially chosen for the shell finite element and repeat the aforementioned procedures and again plot the results against test results or exact closed-form solutions. Plotting the results will determine the best finite element mesh that provides accurate results with less computational time. The studies we conduct to choose the best finite element mesh are commonly called as *convergence studies*. It should be noted that in regions of the structural member where the stress gradient is small, aspect ratios higher than 5 can be used and still can produce satisfactory results. Figure 3.1 shows the finite element mesh used by Ellobody and Young [1.30] to simulate the behavior of fixed-ended rectangular hollow section columns. As mentioned by Ellobody and Young [1.30], "In order to choose the finite element mesh that provides accurate results with minimum computational time, convergence studies were conducted. It is found that the mesh size of 20 mm × 10 mm (length by width) provides adequate accuracy and minimum computational time in modeling the flat portions of cold-formed high strength stainless steel columns, while a finer mesh was used at the corners."

Metal structural members having cross sections that are symmetric about one or two axes can be modeled by cutting half or quarter of the member, respectively, owing to symmetry. Use of symmetry reduces the size of the finite element mesh considerably and consequently reduces the computational time significantly. Detailed discussions on how symmetry can be efficiently used in finite element modeling are presented in Ref. [3.1]. However, researchers and modelers have to be very careful when using symmetry to reduce the mesh size of metal structural members. This is attributed to the fact that most metal columns, beams and beam columns, and connections that have slender cross sections can fail owing to local buckling or local yielding. Failure due to local buckling or local yielding can occur in any region of the metal structural member due to initial local and overall geometric imperfections. Therefore, the whole structural members have to be modeled even if the cross section is symmetric about the two axes. In addition, symmetries have to be in loading, boundary conditions, geometry, and materials. If the cross section is symmetric but the structural member is subjected to different loading along the length of the member or the boundary conditions are not the same at both ends, the whole structural members have to be modeled too. Therefore, it is better to define symmetry in this book as correspondence in size, shape, position of loads, material properties, boundary conditions, residual stresses due to processing, initial local, and overall geometric imperfections that are on opposite sides of a dividing line or plane. As an example, a tensile coupon test specimen can be modeled by considering symmetry, as shown in Figure 3.7. The specimen is an example of a plane-stress uniaxially loaded structure that can be modeled by using triangular and quadrilateral plane-stress solid elements. It can be seen that only quarter of the specimen was modeled due to symmetry. All nodes at symmetry surfaces (1) and (2) were prevented to displace in x-direction and y-direction, respectively. All nodes at the corner location

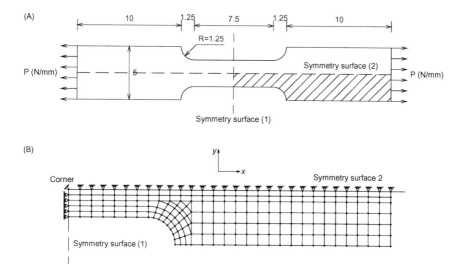

Figure 3.7 Use of symmetry to reduce the size of finite element meshes. (A) Plane-stress uniaxially loaded tensile coupon test specimen. (B) Finite element mesh of quarter of the specimen.

were prevented to displace in x and y-directions. It can also be seen that the mesh is fine at the middle and curved portions of the tensile coupon specimen where stresses are concentrated. The finite element mesh can be coarser at ends where the specimen is fitted in the grips of the tensile testing machine.

It should be noted that most current efficient general-purpose finite element software have the ability to perform meshing of the metal structures automatically. However, in many cases, the resulting finite element meshes may be very fine so that it takes huge time in the analysis process. Therefore, it is recommended in this book to use guided meshing where the modelers apply the aforementioned fundamentals in building the finite element mesh using current software. In this case, automatic meshing software can be of great benefit for modelers.

3.4. MATERIAL MODELING

Most metal structures have nonlinear stress—strain curves or linear—nonlinear stress—strain curves, as shown in Figure 1.1. The stress—strain curves can be determined from tensile coupon tests or stub column tests specified in most current international specifications. The stress—strain curves are characteristic to the construction materials and differ considerably from a material to another. Although the testing procedures of tensile coupon tests and stub column tests are outside the scope of this book, it is important in this chapter to detail how the linear and nonlinear regions of the stress—strain curves are incorporated in the finite element models. The test stress—strain curves obtained from tensile coupon tests are commonly measured with load being applied at a specified loading rate during different time range, which can result in a dynamic stress—strain curve shown in Figure 3.8. The figure was used as an example by Zhu and Young [3.2] in modeling cold-formed steel oval hollow section columns. The nominal (engineering) static stress—strain curve needed for finite element modeling can be obtained from a tensile coupon test by pausing the applied straining for specified few minutes near the proportional limit stress, the yield stress, the ultimate tensile stress, and the post-ultimate tensile stress. This is intended to allow stress relaxation associated with plastic straining to take place. The nominal (engineering) static stress—strain curve is also shown in Figure 3.8. The main important parameters needed from the stress—strain curve are the measured initial Young's modulus (E_0), the measured proportional limit stress (σ_p), the measured static yield stress (σ_y) that is commonly taken as the 0.1% or 0.2% proof stress $(\sigma_{0.1}$ or $\sigma_{0.2})$ for materials having a rounded stress—strain curve with no distinct yield plateau, the measured ultimate tensile strength (σ_u), and the measured elongation after fracture (ε_f). It should be noted that buckling analysis of metal columns, beams and beam columns, and connections commonly involve large inelastic strains. Therefore, the nominal (engineering) static stress—strain curves must be converted to true stress—logarithmic plastic true strain curves. The true

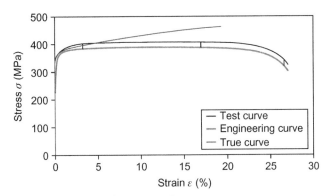

Figure 3.8 Modeling of metal plasticity [3.2].

stress (σ_{true}) and plastic true strain ($\varepsilon_{\text{true}}^{\text{pl}}$) were calculated using Eqs (3.1) and (3.2) as given in ABAQUS [1.27]:

$$\sigma_{\text{true}} = \sigma(1 + \varepsilon) \tag{3.1}$$

$$\varepsilon_{\text{true}}^{\text{pl}} = \ln(1 + \varepsilon) - \sigma_{\text{true}}/E_0 \tag{3.2}$$

where E_0 is the initial Young's modulus, σ and ε are the measured nominal (engineering) stress and strain values, respectively. Figure 3.8 also shows the true stress–plastic true strain curve calculated using Eqs (3.1) and (3.2).

The initial part of the stress–strain curve from origin to the proportional limit stress can be represented based on linear elastic model as given in ABAQUS [1.27]. The linear elastic model can define isotropic, orthotropic, or anisotropic material behavior and is valid for small elastic strains (normally less than 5%). Depending on the number of symmetry planes for the elastic properties, a material can be classified as either isotropic (an infinite number of symmetry planes passing through every point) or anisotropic (no symmetry planes). Some materials have a restricted number of symmetry planes passing through every point; for example, orthotropic materials have two orthogonal symmetry planes for the elastic properties. The number of independent components of the elasticity tensor depends on such symmetry properties. The simplest form of linear elasticity is the isotropic case. The elastic properties are completely defined by giving the Young's modulus (E_0) and the Poisson's ratio (ν). The shear modulus (G) can be expressed in terms of E_0. Values of Poisson's ratio approaching 0.5 result in nearly incompressible behavior.

The nonlinear part of the curve passing the proportional limit stress can be represented based on classical plasticity model as given in ABAQUS [1.27]. The model allows the input of a nonlinear curve by giving tabular values of stresses and strains. When performing an elastic–plastic analysis at finite strains, it is assumed that the plastic strains dominate the deformation and that the elastic strains are small. It is justified because most materials have a well-defined yield stress that is a very small percentage of their Young's modulus. For example, the yield stress of most metals is typically less than 1% of the Young's modulus of the materials. Therefore, the elastic strains will also be less than this percentage, and the elastic response of the materials can be modeled quite accurately as being linear.

The classical metal plasticity models use Mises or Hill yield surfaces with associated plastic flow, which allow for isotropic and anisotropic yield, respectively. The models assume perfect plasticity or isotropic hardening behavior. Perfect plasticity means that the yield stress does not change with plastic strain. Isotropic hardening means that the yield surface changes size uniformly in all directions such that the yield stress increases (or decreases) in all stress directions as plastic straining occurs. Associated plastic flow means that as the material yields, the inelastic deformation rate is in the direction of the normal to the yield surface (the plastic deformation is volume invariant). This assumption is generally acceptable for most calculations with metal. The classical metal plasticity models can be used in any procedure that uses elements with displacement degrees of freedom. The Mises and Hill yield surfaces assume that yielding of the metal is independent of the equivalent pressure stress. The Mises yield surface is used to define isotropic yielding. It is defined by giving the value of the uniaxial yield stress as a function of uniaxial equivalent plastic strain as mentioned previously. The Hill yield surface allows anisotropic yielding to be modeled.

3.5. MODELING OF INITIAL IMPERFECTIONS

Most hot-rolled and cold-formed metal structural members have initial geometric imperfections as a result of the manufacturing, transporting, and handling processes. Initial geometric imperfections can be classified into two main categories, which are local and overall (bow, global, or out-of-straightness) imperfections. Initial local geometric imperfections can be found in any region of the outer or inner surfaces of metal structural members and are in the perpendicular directions to the structural member surfaces. On the other hand, initial overall geometric imperfections are global profiles for the whole structural member along the member length in any direction. Many experimental investigations were presented in the literature highlighting the measurement procedures of initial local and overall geometric imperfections for different structural members, which are outside the scope of this book. However, this book details how the magnitude and profile of initial local and overall geometric imperfections are incorporated in the finite element models. Initial local and overall geometric imperfections can be predicted from finite element models by conducting eigenvalue buckling analysis to obtain the worst cases of local and overall buckling modes.

These local and overall buckling modes can be then factored by measured magnitudes in the tests. Superposition can be used to predict final combined local and overall buckling modes. The resulting combined buckling modes can be then added to the initial coordinates of the structural member. The final coordinates can be used in any subsequent nonlinear analysis. The details of the eigenvalue buckling analysis will be highlighted in Chapter 4.

Accurate finite element models must incorporate initial local and overall geometric imperfections in the analysis; otherwise, the results will not be accurate. Even in most axially loaded metal long column tests, the columns tend to buckle in the direction of the maximum initial overall geometric imperfection. In addition, in most eccentrically loaded metal long column tests, the initial overall geometric imperfection must be added to the eccentricity to obtain the moment resistance of the column. Efficient test programs must include the measurement of initial local and overall geometric imperfections. Figure 3.9 shows the measured initial local geometric imperfection profile of stainless steel rectangular hollow section $200 \times 110 \times 4$ mm as detailed by Young and Lui [1.29]. Table 3.1 shows the measured initial overall geometric imperfections at mid-length of stainless steel columns as detailed by Young and Lui [1.28].

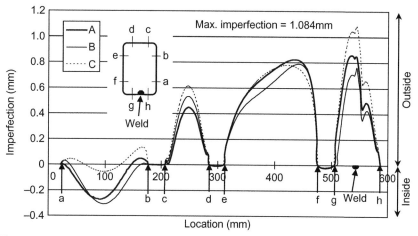

Figure 3.9 Measured local geometric imperfection profiles of RHS $200 \times 110 \times 4$ [1.29].

Table 3.1 Measured Overall Geometric Imperfections at Mid-Length of Columns [1.28]

	δ/L	
Specimen	x	y
SHS1L650	1/430	1/17060
SHS1L1000	1/19685	1/2386
SHS1L1500	1/11811	1/29528
SHS1L2000	1/10499	1/8288
SHS1L2500	1/29528	1/2140
SHS1L3000	1/1390	1/11811
SHS2L650	1/2326	1/4653
SHS2L1000	1/1358	1/2316
SHS2L1500	1/2953	1/1790
SHS2L2000	1/4632	1/1500
SHS2L2500	1/775	1/3076
SHS2L3000	1/872	1/993
RHS1L600	–	1/4295
RHS1L1400	–	1/7349
RHS1L2200	–	1/5588
RHS1L3000	–	1/3236
RHS2L600	–	1/7874
RHS2L1400	–	1/5512
RHS2L2200	–	1/6663
RHS2L3000	–	1/3937

3.6. MODELING OF RESIDUAL STRESSES

Residual stresses are initial stresses existing in cross sections without application of an external load such as stresses resulting from manufacturing processes of metal structural members by cold forming. Residual stresses produce internal membrane forces and bending moments, which are in equilibrium inside the cross sections. The force and the moment resulting from residual stresses in the cross sections must be zero. Residual stresses in structural cross sections are attributed to the uneven cooling of parts of cross sections after hot rolling. Uneven cooling of cross-sectional parts subjects to internal stresses. The parts that cool quicker have residual compressive stresses, while parts that cool lower have residual tensile stresses. Residual stresses cannot be avoided and in most cases are not desirable. The measurement of residual stresses is therefore important for accurate understanding of the performance of metal structural members.

Extensive experimental investigations were conducted in the literature to determine the distribution and magnitude of residual stresses inside cross sections. The experimental investigations can be classified into two main categories: nondestructive and destructive methods. Examples of nondestructive methods are X-ray diffraction and neutron diffraction. Nondestructive methods are suitable for measuring stresses close to the outside surface of cross sections. On the other hand, destructive methods involve machining/cutting of the cross section to release internal stresses and measure resulting change of strains. Destructive methods are based on the destruction of the state of equilibrium of the residual stresses in the cross section. In this way, the residual stresses can be measured by relaxing these stresses. However, it is only possible to measure the consequences of the stress relaxation rather than the relaxation itself. One of the main destructive methods is to cut the cross section into slices and measure the change in strains before and after cutting. After measuring the strains, some simple analytical approaches can be used to evaluate resultant membrane forces and bending moments in the cross sections. Although the testing procedures to determine residual stresses are outside the scope of this book, it is important to detail how to incorporate residual stresses in finite element models. It should be noted that in some cases, incorporating residual stresses can result in small effect on the structural performance of metals. However, in some other cases, it may result in considerable effect. Since the main objective of this book is to accurately model all parameters affecting the behavior and design of metal structures, the way to model residual stresses is highlighted in this book.

Experimental investigations for measuring residual stresses are costly and time consuming. Therefore, some numerical methods were presented in the literature to simulate some typical and simple procedures introducing residual stresses. Dixit and Dixit [3.3] modeled cold rolling for steel and gave a simplified approach to find the longitudinal residual stress. The numerical simulation [3.3] has provided the scope to investigate the effects of different parameters on the magnitude and distribution of residual stresses such as material characteristics and boundary conditions. Kamamato et al. [3.4] have analyzed residual stresses and distortion of large steel shafts due to quenching. The results showed that residual stresses are strongly related to the transformational behavior. Toparli and Aksoy [3.5] analyzed residual stresses during water quenching of cylindrical solid steel bars of various diameters by using finite element technique. The authors have computed the transient temperature distribution for solid bars with general surface heat transfer. Jahanian [3.6] modeled heat treatment and calculated the residual stress in a long solid cylinder by using theoretical and numerical methods with different cooling speeds. Yuan and Wu [3.7] used a finite element program to analyze the transient temperature and residual stress fields for a metal specimen during quenching. They modified the elastic−plastic properties of specimen according to temperature fields. Yamada [3.8] presented a method of solving uncoupled quasi-static thermoplastic problems in perforated plates. In their analysis, a transient thermal stress problem was solved for an infinite plate containing two elliptic holes with prescribed temperature. In all these models, many assumptions were made to simplify the actual process. This is attributed to the fact that it is quite complicated to simulate all the parameters in detail. However, reasonably good models were developed in these pervious investigations by considering the key factors that affect the formation of residual stresses. An extensive survey of the aforementioned numerical investigations was presented by Ding [3.9].

Residual stresses and their distribution are very important factors affecting the strength of axially loaded metal structures. These stresses are of particular importance for slender columns, with slenderness ratio varying from approximately 40 to 120. As a column load is increased, some parts of the column will quickly reach the yield stress and go into the plastic range because of the presence of residual compression stresses. The stiffness will reduce and become a function of the part of the cross section that is still inelastic. A column with residual stresses will behave as though it has a reduced cross section. This reduced cross section or elastic portion of the column will change as the applied load changes. The buckling analysis and post-buckling calculation can be carried out theoretically or numerically by using an effective moment of inertia of the elastic portion of the cross section or by using the tangent modulus. ABAQUS [1.27] is a popular package that can be used for the post-buckling analysis, which gives the history of deflection versus loading. The ultimate strength of the column could be then obtained from this history.

As mentioned previously in Section 1.3 of the book, efficient experimental programs should measure residual stresses in tested specimens, which are detailed as an example in the investigation conducted by Young and Lui [1.28,1.29]. The investigation measured the residual stresses in stainless steel square and rectangular hollow sections. Measurement of residual stresses was carried out using the method of sectioning that requires cutting the hollow section into strips to release the internal residual stresses. The strains before and after cutting were measured by the authors [1.28,1.29]; consequently, residual stresses can be determined. The stainless steel hollow section specimens were marked into strips with an assumed width. A gauge length was marked on the outside and inside mid-surfaces of each strip along the length. The residual strains were measured using an extensometer over the gauge length. The initial readings before cutting were recorded for each strip together with the corresponding temperature. The cutting was carried out using a wire-cutting method in the water to eliminate additional stresses resulting from the cutting process. The readings were taken after cutting and the corresponding temperature was recorded. The readings were corrected for temperature difference before and after cutting. The residual strains were measured for both inner and outer sides of each strip. The membrane residual strain was calculated as the mean of the strains, (inner strain + outer strain)/2. The bending strain was calculated as the difference between the outer and inner strains divided by two, (inner strain − outer strain)/2. A compressive strain (negative value) was recorded at the corner, while a tensile strain (positive value) was recorded at the flat portion. Positive bending strain indicates compressive strain at the inner fiber and tensile strain at the outer fiber. Residual stresses are calculated by multiplying residual strains by Young's modulus of the test specimens. The distribution of membrane and bending residual stresses in the cross section of the test specimen was detailed in [1.28,1.29].

To ensure accurate modeling of the behavior of metal structures, the residual stresses should be included in the finite element models. As an example, the column tests conducted by Young and Lui [1.28,1.29] were modeled by Ellobody and Young [1.30]. Measured residual stresses were implemented in the finite element model as initial stresses using ABAQUS [1.27] software. It should be noted that the slices cut from the cross section to measure the residual stresses can be used to form tensile coupon test specimens. In this case, the effect of bending stresses on the stress−strain curve of the metal material will be considered since the tensile coupon specimen will be tested in the actual bending condition. Therefore, only the membrane residual stresses have to be incorporated in the finite element model as given in Ref. [1.30]. The average values of the measured membrane residual stresses were calculated for corner and flat portions of the section. Figure 1.5 showed an example of the measured membrane residual stresses conducted by Young and Lui [1.28,1.29] and modeled by Ellobody and Young [1.30].

Initial conditions can be specified for particular nodes or elements, as appropriate. The data can be provided directly in an external input file or in some cases by a user subroutine or by the results or output database file from a previous analysis. If initial conditions are not specified, all initial conditions are considered zero in the model. Various types of initial conditions can be specified, depending on the analysis to be performed; however, the type highlighted here is the initial conditions (stresses). The option can be used to apply stresses in different directions. When initial stresses are given, the initial stress state may not be an exact equilibrium state for the finite element model. Therefore, an initial step should be included to check for equilibrium and iterate, if necessary, to achieve equilibrium.

3.7. LOAD APPLICATION

Loads applied on metal structural members in tests or in practice must be simulated accurately in finite element models. Any assumptions or simplifications in actual loads could affect the accuracy of results. There are two common load types applied to metal structural members, which are widely known as *concentrated loads* and *distributed loads*. Concentrated forces and moments can be applied to any node in the finite element model. Concentrated forces and moments are incorporated in the finite element model by specifying nodes, associated degrees of freedom, and magnitude and direction of applied concentrated forces and moments. The concentrated forces and moments could be fixed in direction or alternative can rotate as the node rotates. On the other hand, distributed loads can be prescribed on element faces to simulate surface distributed loads. The application of distributed loads must be incorporated in the finite element model very carefully using appropriate distributed load type that is suitable to each element type. Most software specify different distributed load types associated with the different element types included in the software element library. For example, solid brick elements C3D8 can accept distributed loads on eight surfaces, while shell elements are commonly loaded in planes perpendicular to the shell element mid-surface. Distributed loads can be defined as *element-based* or *surface-based*. Element-based distributed loads can be prescribed on element bodies, element surfaces, or element edges. The surface-based distributed loads can be prescribed directly on geometric surfaces or geometric edges.

Three types of distributed loads can be defined in ABAQUS [1.27], which are body, surface, and edge loads. Distributed body loads are always element-based. Distributed surface loads and distributed edge loads can be element-based or surface-based. Body loads, such as gravity, are applied as element-based loads. The units of body forces are force per unit volume. Body forces can be specified on any elements in the global x-, y-, or z-direction. Also, body forces can be specified on axisymmetric elements in the radial or axial direction. General or shear surface tractions and pressure loads can be applied as element-based or surface-based distributed loads. The units of these loads are force per unit area. Distributed edge tractions (general, shear, normal, or transverse) and edge moments can be applied to shell elements as element-based or surface-based distributed loads. The units of edge tractions are force per unit length. The units of edge moments are torque per unit length. Distributed line loads can be applied to beam elements as element-based distributed loads. The units of line loads are force per unit length. It should be noted that in some cases, distributed surface loads can be transferred to equivalent concentrated nodal loads and can provide reasonable accuracy provided that a fine mesh has been used.

3.8. BOUNDARY CONDITIONS

Following the load application on the finite element model, we can now apply the boundary conditions on the finite element model. Boundary conditions are used in finite element models to specify the values of all basic solution variables such as displacements and rotations at nodes. Boundary conditions can be given as model input data to define zero-valued boundary conditions and can be given as history input data to add, modify, or remove zero-valued or nonzero boundary conditions. Boundary conditions can be specified using either *direct format* or *type format*. The type format is a way of conveniently specifying common types of boundary conditions in stress—displacement analyses. Direct format must be used in all other analysis types. For both direct and type format, the region of the model to which the boundary conditions apply and the degrees of freedom to be restrained must be specified. Boundary conditions prescribed as model data can be modified or removed during analysis steps. In the direct format, the degrees of freedom can be constrained directly in the finite element model by specifying the node number or node set and the degree of freedom to be constrained. As an example in ABAQUS [1.27], when you specify that (CORNER, 1), this means that the node set named (CORNER) are constrained to displace in direction 1 (u_x). While specifying that (CORNER, 1, 4), this means that the node set CORNER are constrained to displace in directions 1—4 $(u_x, u_y, u_z,$ and $\theta_x)$. The type of boundary condition can be specified instead of degrees of freedom. As examples in ABAQUS [1.27], specifying "XSYMM" means symmetry about a plane X = constant, which implies that the degrees of freedom 1, 5, and 6 equal to 0. Similarly, specifying "YSYMM" means symmetry about a plane Y = constant, which implies that the degrees of freedom 2, 4, and 6 equal to 0, and specifying "ZSYMM" means symmetry about a plane Z = constant, which implies that the degrees of freedom 3, 4, and 5 equal to 0. Also, specifying "ENCASTRE" means fully built-in (fixed case), which implies that the degrees of freedom 1, 2, 3, 4, 5, and 6 equal to 0. Finally, specifying "PINNED" means pin-ended case, which implies that the degrees of freedom 1, 2, and 3 equal to 0. Looking again to Figure 3.7, we can now apply a boundary condition of type "XSYMM" to all nodes on symmetry surface (2), and "YSYMM" can be applied to all nodes on symmetry surface (1). It should be noted that once a degree of freedom has been constrained using a type boundary condition as model data, the constraint cannot be modified by using a boundary condition in direct format as model data. Also, a *displacement-type* boundary condition can be used to apply a prescribed displacement magnitude to a degree of freedom.

The application of boundary conditions is very important in finite element modeling. The application must be identical to the actual situation in the metal structural member test or construction. Otherwise, the finite element model will never produce accurate results. Modelers must be very careful in applying all boundary conditions related to the structure and must check that they have not overconstrained the model. Symmetry surfaces also require careful treatment to adjust the boundary conditions at the surface. It should be also noted that applying a boundary condition at a node to constrain this node from displacing or rotating will totally stop this node to displace or rotate. When the displacement or rotation is not completely constrained (partial constraint), springs must be used to apply the boundary conditions with constraint values depending on the stiffness related to the degrees of freedom.

REFERENCES

[3.1] Bowes, W. H. and Russell, L. T. Stress analysis by the finite element method for practicing engineers. Toronto: Lexington Books, 1975.
[3.2] Zhu, J. H. and Young, B. Design of cold-formed steel oval hollow section columns. Journal of Constructional Steel Research, 71, 26—37, 2012.

[3.3] Dixit, U. S. and Dixit, P. M. A study on residual stresses in rolling. International Journal of Machine Tools and Manufacture, 37(6), 837−853, 1997.

[3.4] Kamamato, S., Nihimori, T. and Kinoshita, S. Analysis of residual stress and distortion resulting from quenching in large low-alloy steel shafts. Journal of Materials Sciences and Technology, 1, 798−804, 1985.

[3.5] Toparli, M. and Aksoy, T. Calculation of residual stresses in cylindrical steel bars quenched in water from 600°C, Proceedings of AMSE Conference, vol. 4, New Delhi, India, 93−104, 1991.

[3.6] Jahanian, S. Residual and thermo-elasto-plastic stress distributions in a heat treated solid cylinder. Materials at High Temperatures, 13(2), 103−110, 1995.

[3.7] Yuan, F. R. and Wu, S. L. Transient-temperature and residual-stress fields in axisymmetric metal components after hardening. Journal of Materials Science and Technology, 1, 851−856, 1985.

[3.8] Yamada, K. Transient thermal stresses in an infinite plate with two elliptic holes. Journal of Thermal Stresses, 11, 367−379, 1988.

[3.9] Y. Ding, Residual stresses in hot-rolled solid round steel bars and their effect on the compressive resistance of members. Master Thesis. Windsor (Ontanio, Canada): University of Windsor, 2000.

CHAPTER 4

Linear and Nonlinear Finite Element Analyses

4.1. GENERAL REMARKS

The previous chapter highlighted the main parameters that control finite element modeling of metal structures, and we can now address different linear and nonlinear finite element analyses. This chapter presents the main analyses associated with finite element modeling of metal columns and beams. When the finite element method was introduced in Chapter 2, with a solved example presented in Section 2.7, it was assumed that the displacements of the finite element model are infinitesimally small and that the material is linearly elastic, as shown in Figure 4.1A. In addition, it was assumed that the boundary conditions remain unchanged during the application of loading on the finite element model. With these assumptions, the finite element equilibrium equation was derived for static analysis as presented in Eq. (2.4). The equation corresponds to *linear analysis* of a structural problem because the displacement response $\{d\}$ is a linear function of the applied force vector $\{F\}$. This means that if the forces are increased with a constant factor, the corresponding displacements will be increased with the same factor. On the other hand, in *nonlinear analysis*, the aforementioned assumptions are not valid. The assumption is that the displacement must be small so the evaluation of the stiffness matrix $[K]$ and the force vector $\{F\}$ of Eq. (2.4) were assumed to be constant and independent on the element displacements, because all integrations have been performed over the original volume of the finite elements and the strain–displacement relationships. The assumption of a linear elastic material was implemented in the use of constant stress–strain relationships. Finally, the assumption that the boundary conditions remain unchanged was reflected in the use of constant restraint relations for the equilibrium equation.

Recognizing the previous discussion, we can define three main nonlinear analyses commonly known as materially nonlinear analysis, geometrically (large displacement and large rotation) nonlinear analysis, and materially and geometrically nonlinear analysis. In materially nonlinear

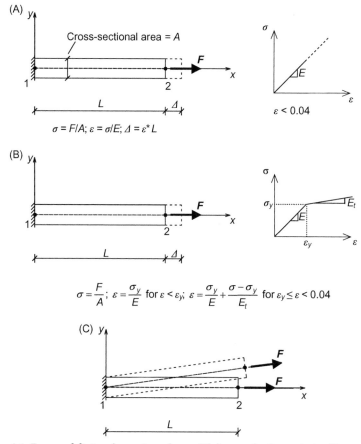

Figure 4.1 Types of finite element analyses: (A) linear elastic analysis, (B) materially nonlinear analysis, and (C) geometrically nonlinear analysis.

analysis, the nonlinear effect lies in the nonlinear stress–strain relationship, with the displacements and strains being infinitesimally small, as shown in Figure 4.1B. Therefore, the usual engineering stress and strain measurements can be employed. In geometrically nonlinear analysis, the structure undergoes large rigid body displacements and rotations, as shown in Figure 4.1C. Majority of geometrically nonlinear analyses were based on von Karman nonlinear equations such as the analyses presented in Refs [4.1–4.8]. The equations allow coupling between bending and membrane behavior with the retention of Kirchhoff normality constraint [1.5]. Finally, materially and geometrically nonlinear analysis combines both nonlinear stress–strain relationship and large displacements and rotations experienced by the structure.

This chapter starts by introducing linear eigenvalue buckling analysis, which is required for modeling initial local and overall geometric imperfections as briefly mentioned in Section 3.5. After that, this chapter details the nonlinear material and geometry analyses required for simulating actual performance of metal structures. Once again, this chapter explains the analyses that are commonly incorporated in all efficient general-purpose finite element software; however as an example, nonlinear analyses used by ABAQUS [1.27] are particularly highlighted. Lastly, this chapter presents the RIKS method used in ABAQUS [1.27] that can accurately model the collapse behavior of metal structural members.

4.2. ANALYSIS PROCEDURES

Most available general-purpose finite element software divides the problem history (overall finite element analysis) into different steps as shown in Figure 4.2. An *analysis procedure* can be specified for each step, with prescribing loads, boundary conditions, and output requests specified for each step. A step is a phase of the problem history, and in its simplest form, a step can be just a static analysis of a load changing from one magnitude to another, as shown in Figure 4.2. For each step, one can choose an analysis procedure. This choice defines the type of analysis to be performed during the step such as static stress analysis, eigenvalue buckling analysis, or any other types of analyses. It should be noted that as mentioned previously,

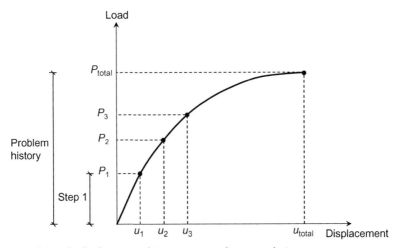

Figure 4.2 Load–displacement history in a nonlinear analysis.

static stress analyses are only detailed in this book. Static analyses are used when inertia effects can be neglected. The analyses can be linear or non-linear and assume that time-dependent material effects, such as creep, are negligible. Linear static analysis involves the specification of load cases and appropriate boundary conditions. If all or part of a structure has linear response, substructuring is a powerful capability for reducing the computational cost of large analyses. Static nonlinear analyses can also involve geometrical nonlinearity and/or material nonlinearity effects. If geometrically nonlinear behavior is expected in a step, the large-displacement formulation should be used. Only one procedure is allowed per step and any combination of available procedures can be used step by step. However, information from a previous step can be imported to the current step by calling the results from the previous step. The loads, boundary conditions, and output requests can be inserted in any step.

Most available general-purpose finite element software classify the steps into two main kinds of steps: *general analysis steps* and *linear perturbation steps*. General analysis steps can be used to analyze linear or nonlinear response. On the other hand, linear perturbation steps can be used only to analyze linear problems. Linear analysis is always considered to be linear perturbation analysis about the state at the time when the linear analysis procedure is introduced. The linear perturbation approach allows general application of linear analysis techniques in cases where the linear response depends on preloading or on the nonlinear response history of the model. In general analysis steps and linear perturbation steps, the solution to a single set of applied loads can be predicted. However, for static analyses covered in this book, it is possible to find solutions to multiple load cases. In this case, the overall analysis procedure can be changed from step to step. This allows the state of the model (stresses, strains, displacements, deformed shapes, etc.) to be updated throughout all general analysis steps. The effects of previous history can be included in the response in each new analysis step by calling the results of a previous history. As an example, after conducting an initial condition analysis step to include residual stresses in cross sections, the initial stresses in the whole cross section will be updated from zero to new applied stresses that accounted for the residual stresses effect in metal structures.

It should be noted that linear perturbation steps have no effect on subsequent general analysis steps and can be conducted separately as a whole (overall) analysis procedure. In this case, the data obtained from the linear perturbation steps can be saved in files that can be called into the subsequent general analysis steps. For example, linear eigenvalue buckling analyses, needed for modeling of initial overall and local geometric imperfections, can be conducted initially as a separate overall analysis procedure, and buckling modes can be extracted from the analyses and saved in files. The saved files can be called into subsequent static general analyses and factored to model initial geometric imperfections. The most obvious reason for using several steps in an analysis is to change the analysis procedure type. However, several steps can also be used to change output requests, such as the boundary conditions or loading (any information specified as history, or step-dependent data). Sometimes, an analysis may be progressed to a point where the present step definition needs to be modified. ABAQUS [1.27] provides the ability to restart the analysis, whereby a step can be terminated prematurely and a new step can be defined for the problem continuation. History data prescribing the loading, boundary conditions, and output will remain in effect for all subsequent general analysis steps until they are modified or reset. ABAQUS [1.27] will compare all loads and boundary conditions specified in a step with the loads and boundary conditions in effect during the previous step to ensure consistency and continuity. This comparison is expensive if the number of individually specified loads and boundary conditions is very large. Hence, the number of individually specified loads and boundary conditions should be minimized, which can usually be done by using element and node sets instead of individual elements and nodes.

Most current general-purpose finite element software divides each step of analysis into multiple increments. In most cases, one can choose either *automatic (direct) time incrementation* or *user-specified fixed time incrementation* to control the solution. Automatic time incrementation is a built-in incrementation scheme that allows the software to judge the increment needed based on equilibrium requirements. On the other hand, user-specified fixed time incrementation forces the software to use a specified fixed increment, which in many cases may be large, small, or need updating during the step. This results in the analysis to be stopped and readjusted. Therefore, automatic incrementation is recommended for most cases. The methods for selecting automatic or direct incrementation are always prescribed by all general-purpose software to help modelers. In nonlinear analyses, most general-purpose software will use increment and iterate as necessary to analyze a step, depending on the severity of the nonlinearity. Iterations conducted within an increment can be classified as *regular equilibrium iterations* and *severe discontinuity iterations*. In regular equilibrium iterations, the solution varies smoothly, while in severe discontinuity iterations abrupt changes in stiffness occur. The analysis will continue to iterate until the severe discontinuities are sufficiently small (or no severe discontinuities occur) and the equilibrium tolerances are satisfied. Modelers can provide parameters to indicate a level of accuracy in the time integration, and the software will choose the time increments to achieve this accuracy. Direct user control is provided because it can sometimes save computational cost in cases where modelers are familiar with the problem and know a suitable incrementation scheme. Modelers can define the upper limit to the number of increments in an analysis. The analysis will stop if this maximum is exceeded before the complete solution to the step has been obtained. To reach a solution, it is often necessary to increase the number of increments allowed by defining a new upper limit.

In nonlinear analyses, general-purpose software use *extrapolation* to speed up the solution. Extrapolation refers to the method used to determine the first guess to the incremental solution. The guess is determined by the size of the current time increment and by whether *linear, parabolic,* or no extrapolation of the previously attained history of each solution variable is chosen. Linear extrapolation is commonly used with 100% extrapolation of the previous incremental solution being used at the start of each increment to begin the nonlinear equation solution for the next increment. No extrapolation is used in the first increment of a step. Parabolic extrapolation uses two previous incremental solutions to obtain the first guess to the current incremental solution. This type of extrapolation is useful in situations when the local variation of the solution with respect to the timescale of the problem is expected to be quadratic, such as the large rotation of structures. If parabolic extrapolation is used in a step, it begins after the second increment of the step, i.e., the first increment employs no extrapolation, and the second increment employs linear extrapolation. Consequently, slower convergence rates may occur during the first two increments of the succeeding steps in a multistep analysis. Nonlinear problems are commonly solved using Newton's method, and linear problems are commonly solved using the stiffness method. Details of the aforementioned solution methods are outside the scope of this book; however, the methods are presented in detail in Refs [1.1−1.7].

Most general-purpose software adopt a *convergence criterion* for the solution to nonlinear problems automatically. Convergence criterion is the method used by software to govern the balance equations during the iterative solution. The iterative solution is commonly used to solve the equations of nonlinear problems for unknowns, which are the degrees of freedom at the nodes of the finite element model. Most general-purpose software have control parameters designed to provide reasonably optimal solution to complex problems involving combinations of nonlinearities as well as efficient solution to simpler nonlinear cases. However, the most important consideration in the choice of the control parameters is that any solution accepted as "converged" is a close approximation to the exact solution to the nonlinear equations. Modelers can reset many solution control parameters related to the tolerances used for equilibrium equations. If less strict convergence criterion is used, results may be accepted as converged when they are not sufficiently close to the exact solution to the nonlinear equations. Caution should be considered when resetting solution control parameters. Lack of convergence is often due to modeling issues, which should be resolved before changing the accuracy controls. The solution can be terminated if the balance equations failed to converge. It should be noted that linear cases do not require more than one equilibrium iteration per increment, which is easy to converge. Each increment of a nonlinear solution will usually be solved by multiple equilibrium iterations. The number of iterations may become excessive, in which case the increment size should be reduced and the increment will be attempted again. On the other hand, if successive increments are solved with a minimum number of iterations, the increment size may be increased. Modelers can specify a number of time incrementation control parameters. Most general-purpose software may have trouble with the element calculations because of excessive distortion in large-displacement problems or because of very large plastic strain increments. If this occurs and automatic time incrementation has been chosen, the increment will be attempted again with smaller time increments.

4.3. LINEAR EIGENVALUE BUCKLING ANALYSIS

Eigenvalue buckling analysis is generally used to estimate the critical buckling (bifurcation) load of structures. The analysis is a linear perturbation procedure. The analysis can be the first step in a global analysis of an unloaded structure or it can be performed after the structure has been preloaded. It can be used to model measured initial overall and local geometric imperfections or in the investigation of the imperfection sensitivity of a structure in case of lack of measurements. Eigenvalue buckling is generally used to estimate the critical buckling loads of *stiff* structures (classical eigenvalue buckling). Stiff structures carry their design loads primarily by axial or membrane action, rather than by bending action. Their response usually involves very little deformation prior to buckling. A simple example of a stiff structure is the stainless steel hollow section columns presented in Figure 3.1, which responds very stiffly to a compressive axial load until a critical load is reached, when it bends suddenly and exhibits a much lower stiffness. However, even when the response of a structure is nonlinear before collapse, a general eigenvalue buckling analysis can provide useful estimates of collapse mode shapes.

The buckling loads are calculated relative to the original state of the structure. If the eigenvalue buckling procedure is the first step in an analysis, the buckled (deformed) state of the model at the end of the eigenvalue buckling analysis step will be the updated original state of the structure. The eigenvalue buckling can include preloads such as dead load and other loads. The preloads are often zero in classical eigenvalue buckling analyses. An incremental loading pattern is defined in the eigenvalue buckling prediction step. The magnitude of this loading is not important; it will be scaled by the load multipliers that are predicted by the eigenvalue buckling analysis. The buckling mode shapes (eigenvectors) are also predicted by the eigenvalue buckling analysis. The critical buckling loads are then equal to the preloads plus the scaled incremental load. Normally, the lowest load multiplier and buckling mode are of interest. The buckling mode shapes are normalized vectors and do not represent actual magnitudes of deformation at critical load. They are normalized so that the maximum displacement component has a magnitude of 1.0. If all displacement components are zero, the maximum rotation component is normalized to 1.0. These buckling mode shapes are often the most useful outcome of the eigenvalue buckling analysis, since they predict the likely failure modes of the structure.

Some structures have many buckling modes with closely spaced eigenvalues, which can cause numerical problems. In these cases, it is recommended to apply enough preload to load the structure to just below the buckling load before performing the eigenvalue analysis. In many cases, a series of closely spaced eigenvalues indicates that the structure is imperfection sensitive. An eigenvalue buckling analysis will not give accurate predictions of the buckling load for imperfection-sensitive structures. In this case, the static Riks procedure, used by ABAQUS [1.27], which will be highlighted in this chapter, should be used instead. Negative eigenvalues may be predicted from an eigenvalue buckling analysis. The negative eigenvalues indicate that the structure would buckle if the loads were applied in the opposite direction. Negative eigenvalues may correspond to buckling modes that cannot be understood readily in terms of physical behavior, particularly if a preload is applied that causes significant geometric nonlinearity. In this case, a geometrically nonlinear load—displacement analysis should be performed. Because buckling analysis is usually done for stiff structures, it is not usually necessary to include the effects of geometry change in establishing equilibrium for the original state. However, if significant geometry change is involved in the original state and this effect is considered to be important, it can be included by specifying that geometric nonlinearity should be considered for the original state step. In such cases, it is probably more realistic to perform a geometrically nonlinear load—displacement analysis (Riks analysis) to determine the collapse loads, especially for imperfection-sensitive structures as mentioned previously. While large deformation can be included in the preload, the eigenvalue buckling theory relies on there being little geometry change due to the *live* (scaled incremental load) buckling load. If the live load produces significant geometry changes, a nonlinear collapse (Riks) analysis must be used.

The initial conditions such as residual stresses can be specified for an eigenvalue buckling analysis. If the buckling step is the first step in the analysis, these initial conditions form the original state of the structure. Boundary conditions can be applied to any of the displacement or rotation degrees of freedom (six degrees of freedom). Boundary conditions are treated as constraints during the eigenvalue buckling analysis. Therefore, the buckling mode shapes are affected by these boundary conditions. The buckling mode shapes of symmetric structures subjected to symmetric loadings are either symmetric or antisymmetric. In such cases, it is more efficient to use symmetry to reduce the finite element mesh of the model. Axisymmetric structures subjected to compressive loading often collapse in nonaxisymmetric modes. Therefore, these structures must be modeled as a whole. The loads prescribed in an eigenvalue buckling analysis can be concentrated nodal forces applied to the displacement degrees of freedom or can be distributed loads applied to finite element faces. The load stiffness can be of a significant effect on the critical buckling load. It is important that the structure is not preloaded above the critical buckling load. During an eigenvalue buckling analysis, the model's response is defined by its linear elastic stiffness in the original state. All nonlinear or inelastic material properties are ignored during an eigenvalue buckling analysis. Any structural finite elements can be used in an eigenvalue buckling analysis. The values of the eigenvalue load multiplier (buckling loads) will be printed in the data files after the eigenvalue buckling analysis. The buckling mode shapes can be visualized using the software. Any other information such as values of stresses, strains, or displacements can be saved in files at the end of the analysis.

Now, let us go back to the fixed-ended cold-formed stainless steel rectangular hollow section column presented in Figure 3.1. It is possible to explain how the eigenvalue buckling analysis has been used to model initial overall and local geometric imperfections of stainless steel columns presented by Ellobody and Young [1.30]. As mentioned previously, long columns having compact cross sections with small overall depth to plate thickness ratios (D/t) are likely to fail owing to overall flexural buckling. On the other hand, long columns having slender or relatively slender cross sections with large D/t ratios are likely to fail due to local buckling or interaction of local and overall buckling. Both initial local and overall geometric imperfections were found in the columns as a result of the manufacturing, transporting, and fitting processes. Hence, superposition of local buckling mode as well as overall buckling mode with measured magnitudes is recommended [4.9,4.10] in the finite element modeling of the column. These buckling modes can be obtained by carrying eigenvalue analyses of the column with large D/t ratio as well as small D/t ratio to ensure local and overall buckling occurs, respectively. In this case, only the lowest buckling mode (eigenmode 1) was used in the eigenvalue buckling analyses. This technique is used in this study to model the initial local and overall imperfections of the columns. Slender stub columns having short lengths can be modeled for local imperfection only without the consideration of overall imperfection. Since all buckling modes predicted by ABAQUS [1.27] eigenvalue analysis are normalized to 1.0, the buckling modes were factored by the measured magnitudes of the initial local and overall geometric imperfections. Figure 4.3 shows the local and overall imperfection buckling modes predicted for the column presented in Figure 3.1.

4.4. MATERIALLY NONLINEAR ANALYSIS

Materially nonlinear analysis of metal structures is a general nonlinear analysis step. The analysis can be also called *load–displacement nonlinear*

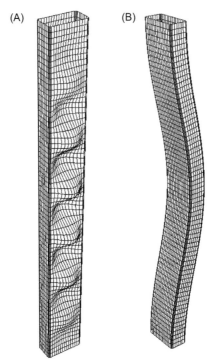

Figure 4.3 Buckling modes (eigenmode 1) for the column specimen presented in Figure 3.1 and previously reported in Ref. [1.30]: (A) local imperfection and (B) overall imperfection.

material analysis and normally follows the linear eigenvalue buckling analysis step or initial condition stress analysis. All required information regarding the behavior of metal structures are predicted from the materially nonlinear analysis. The information comprised the ultimate loads, failure modes, and load—displacement relationships as well as any other required data can be obtained from materially nonlinear analysis. The initial overall and local geometric imperfections, residual stresses, and nonlinear stress—strain curves of the construction material are included in the load—displacement nonlinear material analysis. Since most, if not all, metal structures have nonlinear stress—strain curves or linear—nonlinear stress—strain curves, which are shown for examples in Figure 1.1, most of the general nonlinear analysis steps associated with metal structures are materially nonlinear analyses. Section 3.4 has previously detailed the modeling of nonlinear material properties that should be included in the materially nonlinear analyses.

Materially nonlinear analysis (with or without consideration of geometric nonlinearity) of metal structures is done to determine the overall response of the structures. From a numerical viewpoint, the implementation of a nonlinear stress–strain curve of a construction metal material involves the integration of the state of the material at an integration point over a time increment during a materially nonlinear analysis. The implementation of a nonlinear stress–strain curve must provide an accurate material stiffness matrix for use in forming the nonlinear equilibrium equations of the finite element formulation. The mechanical constitutive models associated with metal structures in ABAQUS [1.27] consider elastic and inelastic response of the material. The inelastic response is commonly modeled with plasticity models as mentioned previously in Chapter 3. In the inelastic response models that are provided in ABAQUS [1.27], the elastic and inelastic responses are distinguished by separating the deformation into *recoverable* (elastic) and *nonrecoverable* (inelastic) parts. This separation is based on the assumption that there is an additive relationship between strain rates of the elastic and inelastic parts. The constitutive material models used in most available general-purpose finite element software are commonly accessed by any of the solid or structural elements previously highlighted in Chapters 2 and 3. This access is made independently at each constitutive calculation point. These points are the numerical integration points in the elements. The constitutive models obtain the state at the point under consideration at the start of the increment from the material database specified in the step. The state variables include the stresses and strains used in the constitutive models. The constitutive models update the state of the material response to the end of the increment. Some examples of materially nonlinear analyses are presented in Refs [4.11,4.12].

4.5. GEOMETRICALLY NONLINEAR ANALYSIS

Geometrically nonlinear analysis of metal structures is a general nonlinear analysis step. The analysis can be also called *load–displacement nonlinear geometry analysis* and normally follows the linear eigenvalue buckling analysis step or initial condition stress analysis. The initial overall and local geometric imperfections and residual stresses are included in the load–displacement nonlinear geometry analysis. If the stress–strain curve of the construction metal material is nonlinear, the analysis will be called *combined materially and geometrically nonlinear analysis* or *load–displacement nonlinear material and geometry analysis*, as shown for examples in Refs [4.13,4.14]. All required information regarding the behavior of metal

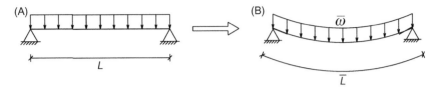

Figure 4.4 Simply supported beam in a geometrically nonlinear analysis: (A) bending moments expected only and (B) membrane forces and bending moments expected.

structures are predicted from the combined materially and geometrically nonlinear analysis. The information comprised the ultimate loads, failure modes, and load—displacement relationships as well as any other required data can be obtained from the combined materially and geometrically nonlinear analysis.

In order to understand the geometrically nonlinear analysis, let us imagine a simply supported beam subjected to lateral loads producing only bending moments at small loads, as shown in Figure 4.4A. As deflections increase at higher loads, there will be membrane forces in addition to bending moments. In this case, large displacements and rotations may constitute a major part of the overall motion of the beam. If the lateral deflection increases significantly, the classical theory of beams will be inadequate and the second-order effect of the vertical displacements on the membrane stresses needs to be considered. In addition, all classical solutions for elastic beams will not be applicable to beams loaded beyond the elastic limit. Further details in geometrically nonlinear analyses could be found in Refs [1.1—1.7].

4.6. RIKS METHOD

The Riks method provided by ABAQUS [1.27] is an efficient method that is generally used to predict unstable, geometrically nonlinear collapse of a structure. The method can include nonlinear materials and boundary conditions. The method commonly follows an eigenvalue buckling analysis to provide complete information about a structure's collapse. The Riks method can be used to speed convergence of unstable collapse of structures. Geometrically nonlinear static metal structures sometimes involve buckling or collapse behavior. Several approaches are possible for modeling such behavior. One of the approaches is to treat the buckling response dynamically, thus actually modeling the response with inertia effects

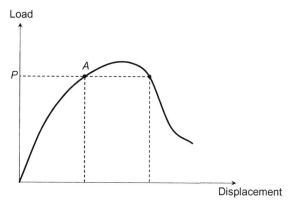

Figure 4.5 Load–displacement behavior that could be predicted by the Riks method (ABAQUS [1.27]).

included as the structure snaps. This approach is easily accomplished by restarting the terminated static procedure and switching to a dynamic procedure when the static solution becomes unstable. In some simple cases, displacement control can provide a solution, even when the conjugate load (the reaction force) is decreasing as the displacement increases. Alternatively, static equilibrium states during the unstable phase of the response can be found by using the *modified Riks method* supported by ABAQUS [1.27]. This method is used for cases where the loading is proportional, where the load magnitudes are governed by a single scalar parameter. The method can provide solutions even in cases of complex, unstable response such as that shown in Figure 4.5.

In simple structures, linear eigenvalue buckling analysis may be sufficient for design evaluation. However, in complex structures involving material nonlinearity, geometric nonlinearity prior to buckling, or unstable postbuckling behavior, a load–displacement (Riks) analysis must be performed to investigate the structures accurately. The Riks method treats the load magnitude as an additional unknown and solves loads and displacements simultaneously. Therefore, another quantity must be used to measure the progress of the solution. ABAQUS [1.27] uses the arc length along the static equilibrium path in load–displacement domain. This approach provides solutions regardless of whether the response is stable or unstable. If the Riks step is a continuation of a previous history, any loads that exist at the beginning of the step are treated as dead loads with constant magnitude. A load whose magnitude is defined in the Riks step is referred to as a reference load. All prescribed loads are ramped from the initial (dead load) value to the reference values specified. ABAQUS [1.27] uses Newton's method to solve the nonlinear equilibrium equations. The Riks procedure uses very small extrapolation of the strain increment. Modelers can provide an initial increment in arc length along the static equilibrium path when defining the step. After that, ABAQUS [1.27] computes subsequent steps automatically. Since the loading magnitude is part of the solution, modelers need a method to specify when the step is completed. It is common that one can specify a maximum displacement value at a specified degree of freedom. The step will terminate once the maximum value is reached. Otherwise, the analysis will continue until the maximum number of increments specified in the step definition is reached.

The Riks method works well with structures having a smooth equilibrium path in load—displacement domain. The Riks method can be used to solve postbuckling problems, both with stable and unstable postbuckling behavior. In this way, the Riks method can be used to perform postbuckling analyses of structures that show linear behavior prior to (bifurcation) buckling. When performing a load—displacement analysis using the Riks method, important nonlinear effects can be included. Imperfections based on linear buckling modes can be also included in the analysis of structures using the Riks method. It should be noted that the Riks method cannot obtain a solution at a given load or displacement value since these are treated as unknowns. Termination of the analysis using the Riks method occurs at the first solution that satisfies the step termination criterion. To obtain solutions at exact values of load or displacement, the analysis must be restarted at the desired point in the step and a new, non-Riks step must be defined. Since the subsequent step is a continuation of the Riks analysis, the load magnitude in that step must be given appropriately so that the step begins with the loading continuing to increase or decrease according to its behavior at the point of restart. Initial values of stresses such as residual stresses can be inserted in the analysis using the Riks method. Also, boundary conditions can be applied to any of the displacement or rotation degrees of freedom (six degrees of freedom). Concentrated nodal forces and moments applied to associated displacement or rotation degrees of freedom (six degrees of freedom) as well as distributed loads at finite element faces can be inserted in the analysis using the Riks method. Nonlinear material models that describe mechanical behavior of metal structures can be incorporated in the analysis using the Riks method.

REFERENCES

[4.1] Murray, D. W. and Wilson, E. L. Finite element large deflection analysis of plates. American Society of Civil Engineers, ASCE, 95(1), 143−165, 1969.

[4.2] Bergan, P. G. and Clough, R. W. Large deflection analysis of plates and shallow shells using the finite element method. International Journal for Numerical Methods in Engineering, 5, 543−556, 1973.

[4.3] Srinivasan, R. S. and Bobby, W. Nonlinear analysis of skew plates using the finite element method. Composite Structures, 6, 199−202, 1976.

[4.4] Argyris, J. H., Dunne, P. C. and Scharpf, D. W. On large displacement-small strain analysis of structures with rotational freedoms. Computer Methods in Applied Mechanics and Engineering, 14, 401−451, 1978.

[4.5] Kondoh, K., Tanaka, K. and Atluri, S. N. An explicit expression for the tangent-stiffness of a finitely deformed 3-D beam and its use in the analysis of space frames. Composite Structures, 24, 253−271, 1986.

[4.6] Yang, Y. B. and McGuire, W. Joint rotation and geometric nonlinear analysis. Journal of Structural Engineering, ASCE, 112, 879−905, 1986.

[4.7] Meek, J. L. and Loganathan, S. Geometrically non-linear behaviour of space frame structures. Composite Structures, 31, 35−45, 1989.

[4.8] Surana, K. S. and Sorem, R. M. Geometrically non-linear formulation for three dimensional curved beam elements with large rotations. International Journal for Numerical Methods in Engineering, 28, 43−73, 1989.

[4.9] Ellobody, E. and Young, B. Behaviour of cold-formed steel plain angle columns. Journal of Structural Engineering, ASCE, 131(3), 457−466, 2005.

[4.10] Young, B. and Ellobody, E. Buckling analysis of cold-formed steel lipped angle columns. Journal of Structural Engineering, ASCE, 131(10), 1570−1579, 2005.

[4.11] Holzer, S. and Yoshbash, Z. The p-version of the finite element method in incremental elasto-plastic analysis. International Journal for Numerical Methods in Engineering, 39, 1859−1878, 1996.

[4.12] Duster, A. and Rank, E. The p-version of the finite element method compared to an adaptive h-version for the deformation theory of plasticity. Computer Methods in Applied Mechanics and Engineering, 190, 1925−1935, 2001.

[4.13] Kassimali, A. Large deformation analysis of elastic-plastic frames. Journal of Structural Engineering, 109, 1869−1886, 1983.

[4.14] Nie, Q. and Niu, Q. p-Version large strain finite element formulation and application in elastic-plastic deformation. Composite Structures, 65, 761−765, 1997.

CHAPTER 5

Examples of Finite Element Models of Metal Columns

5.1. GENERAL REMARKS

The insight of the main analysis procedures associated with finite element modeling of metal structures has been provided in earlier chapters; we can now present some examples of different finite element models of metal columns. The examples presented in this chapter have been published in journal papers that successfully detailed the performance of metal columns. It should be noted that the examples presented in this chapter are arbitrarily chosen from the research conducted by the authors of this book so that all related information regarding the finite element models developed in the papers can be available to readers. The chosen finite element models are columns constructed from different metals having different mechanical properties, different cross sections, different boundary conditions, and different geometries. It should be also noted that when presenting the previously published models, the main objective is not to repeat the previous published information but to explain the fundamentals of the finite element method used in developing the models.

This chapter first presents a survey of recently published numerical, using finite element method, investigations on metal columns. After that, the chapter presents four examples of finite element models previously published by the authors for four different metal columns. The authors will highlight in this chapter how the information presented in the previous chapters are used to develop the examples of finite element models discussed here. The experimental investigations were simulated using the developed finite element models based on the information provided in this book. The finite element models were verified, and the results were compared with design values calculated from current specifications are presented in this chapter with clear references. The authors have an aim that the presented examples highlighted in this chapter can explain to readers the effectiveness of finite element models in providing detailed data that augment experimental investigations conducted in the field.

Finite Element Analysis and Design of Metal Structures
DOI: http://dx.doi.org/10.1016/B978-0-12-416561-8.00005-6

The results are discussed to show the significance of the finite element models in predicting the structural response of different metal structural elements.

5.2. PREVIOUS WORK

Many finite element models were developed in the literature, with detailed examples presented in the following paragraphs, to investigate the behavior and design of metal columns. The aforementioned numerical investigations were performed on metal columns of steel, cold-formed steel, stainless steel, and aluminum alloy materials. Schmidt [5.1] has presented a state-of-the-art review of available information regarding the stability and design of steel shell structures. The review has focused on the various approaches related to a numerical-based stability design. The study [5.1] has credited the finite element analyses of stability problems of metal structures detailed by Galambos [5.2]. The author has outlined that the development of powerful computers and highly efficient numerical techniques helped the calculation related to shell structures with complicated geometry, dominant imperfection influences, and nonlinear load carrying behavior. The study [5.1] has also highlighted that, in the last decade, necessary numerical tools are in the hands of research academics as well as in commercial finite element software such as ABAQUS [1.27] and ANSYS [5.3] for ordinary structural design engineers. The study [5.1] has also shown that the main task of the design engineer is to model complicated metal structures or metal shells properly and to convert the numerical output into the characteristic buckling strength of real shells, which is needed for an equally safe and economic design as recommended by Schmidt and Krysik [5.4]. Furthermore, the study [5.1] has shown that numerical investigations have been incorporated in relevant guidance in the draft of the European Code (EC3) BS EN 1993-1-6 [5.5] for steel shell structures. The study [5.1] has discussed different numerical approaches to shell buckling using commercial finite element general-purpose software. The author has concluded the need for shell buckling tests, with high quality to be used as physical verification benchmarks for numerical models.

Narayanan and Mahendran [5.6] have detailed combined experimental and numerical investigations highlighting the distortional buckling behavior of different shapes of cold-formed steel columns. The authors have carried out more than 15 tests on the columns having intermediate length under axial compression for the verification of the finite element models. The authors have determined the sections and buckling properties of the columns using a finite strip analysis program called THIN-WALL developed by the University of Sydney. The numerical investigation performed in the study [5.6] used the general-purpose software ABAQUS [1.27]. The developed finite element models have incorporated initial geometric imperfections and residual stresses. The load—axial shortening and load—strain relationships were predicted from the finite element models and compared against test results. The authors have compared the ultimate design load capacities predicted from the tests as well as finite element analyses against the design strengths calculated using the Australian/New Zealand Standard (AS/NZS) 4600 [5.7]. The study has used the developed finite element models to perform parametric studies considering different steel strengths, thicknesses, and column lengths. The 4-node 3D quadrilateral shell elements with reduced integration (S4R5), refer to Section 3.2, available in ABAQUS [1.27] library were used in the finite element analyses performed in the study [5.6]. An eigenvalue buckling analysis, as detailed in Section 4.3, was performed first to obtain the buckling loads and associated buckling modes. The authors have performed a series of convergence studies, as shown in Section 3.3, to predict the reasonable finite element mesh size, which was 5×5 mm. Elastic-perfectly plastic material properties were assumed for all steel grades used in the tests and finite element analyses. In the nonlinear analyses [5.6], initial local geometric imperfections were modeled by providing initial out-of-plane deflections to the model. The first elastic buckling mode shape was used to create the local geometric imperfections for the nonlinear analysis. The maximum amplitude of the buckled shape determined the degree of imperfection. The maximum value of distortional imperfection was taken based on the recommendations by Schafer and Peköz [5.7] and Kwon and Hancock [5.8]. The study [5.6] did not include any overall geometric imperfections in the finite element analyses. The residual stresses were modeled using the INITIAL CONDITIONS option with TYPE = STRESS, USER available in ABAQUS [1.27], as detailed in Section 3.6. The user-defined initial stresses were created using the SIGINI Fortran user subroutine.

Raftoyiannis and Ermopoulos [5.9] have studied the elastic stability of eccentrically loaded steel columns with tapered and stepped cross sections. The initial geometric imperfection was incorporated in the analysis assuming a parabolic shape according to EC3 [5.10]. A nonlinear finite element analysis was employed by the authors to predict the plastic loads and buckling behavior of the columns. The nonlinear analyses were performed using a finite element package [5.11]. The flanges and web of the steel columns were modeled with flat quadrilateral 3D shell elements. An incremental procedure was employed for the applied load until a failure mode was reached. The geometrical nonlinearity with large displacements was considered in the finite element analyses using the updated Lagrange method for solution to the nonlinear problem [1.1−1.7].

Zhu and Young [5.12] have presented a numerical investigation on fixed-ended aluminum alloy tubular columns of square and rectangular hollow sections. The columns investigated [5.12] were fixed-ended columns with both ends transversely welded to aluminum end plates. The failure modes predicted from the finite element analyses comprised local buckling, flexural buckling, and interaction of local and flexural buckling. The initial local and overall geometric imperfections were incorporated in the finite element analyses as detailed in Ref. [5.12]. The material nonlinearity of aluminum alloy was considered in the analysis. The load-shortening curves predicted by the finite element analysis were compared against test results. The 4-node doubly curved shell elements with reduced integration (S4R) were used in the model based on the previous recommendations given by Yan and Young [5.13] and Ellobody and Young [1.30]. The size of the finite element mesh used in [5.12] was 10×10 mm (length by width), which was used in the modeling of the columns. The same size has been previously used to simulate axially loaded fixed-ended columns and shown to provide good simulation results [5.13]. Both initial local and overall geometric imperfections were incorporated in the model. Superposition of local buckling mode and overall buckling mode with the measured magnitudes was carried out, as discussed in Section 3.5. The buckling modes were obtained by eigenvalue buckling analysis of the columns with very high value of width-to-thickness ratio and very low value of width-to-thickness ratio to ensure local and overall buckling occurs, respectively. Only the lowest buckling mode (eigenmode 1) was used in the eigenvalue analysis. Residual stresses were not included in the finite element model [5.12] because in extruded aluminum alloy profiles, residual stresses are small and can be neglected as recommended in Ref. [5.14].

The structural performance of cold-formed stainless steel slender and nonslender circular hollow section columns was previously investigated by Young and Ellobody [5.15] and Ellobody and Young [5.16], respectively, through numerical investigations. Nonlinear 3D finite element models were developed by the authors highlighting the behavior of the normal strength austenitic stainless steel type 304 and the high strength duplex (austenitic-ferritic approximately equivalent to EN 1.4462 and UNS S31803) columns. The columns were compressed between fixed ends at different column lengths. The geometric and material nonlinearities have been included in the finite element analysis. The column strengths and failure modes were predicted. An extensive parametric study was carried out to study the effects of normal and high strength materials on cold-formed stainless steel non-slender circular hollow section columns. The column strengths predicted from the finite element analysis were compared with the design strengths calculated using the American Specification [5.17], Australian/New Zealand Standard [5.18], and European Code [5.19] for cold-formed stainless steel structures. The numerical investigations [5.15, 5.16] have proposed improved design equations for cold-formed stainless steel slender and nonslender circular hollow section columns, respectively. The finite element analyses performed in Refs [15.15, 15.16] have used ABAQUS [1.27]. The finite element models were verified against the tests conducted by Young and Hartono [5.20] on cold-formed stainless steel circular hollow section columns.

The developed finite element models used the measured geometry, initial local, and overall geometric imperfections and material properties. The 4-node doubly curved shell elements with reduced integration (S4R) was used to model the buckling behavior of fixed-ended cold-formed stainless steel circular hollow section columns. The mesh size used in the model was approximately 10×10 mm (length by width). The load was applied in increments using the modified Riks method, as detailed in Section, available in the ABAQUS [1.27] library. The nonlinear geometry parameter (*NLGEOM) was included to deal with the large displacement analysis. The load application and boundary conditions were identical to the tests [5.20]. The measured stress—strain curves of circular stainless steel tubes [5.20] were used in the analysis. Both initial local and overall geometric imperfections were found in the column specimens. Hence, superposition of local buckling mode as well as overall buckling mode with measured magnitudes was used in the finite element analysis. These buckling modes were obtained by carrying eigenvalue analyses of the column with large external diameter to plate thickness ratio (D/t) as well as small D/t ratio to ensure local and overall buckling occurs, respectively. Only the lowest buckling mode (eigenmode 1) was obtained from the eigenvalue analyses. Previous studies by Gardner [5.21], and Ellobody and

Young [1.30] on cold-formed stainless steel square and rectangular hollow section columns have shown that the effect of residual stresses on the column ultimate load is considered to be quite small. The cold-formed square hollow section is formed by cold-rolling with welds of annealed flat strip into a circular hollow section, and then further rolled into square hollow section. Hence, the effect of residual stresses on the strength and behavior of cold-formed stainless steel circular hollow section columns would be even smaller than the square and rectangular hollow section columns. Therefore, in order to avoid the complexity of the analysis, the residual stresses were not included in the finite element analysis of cold-formed stainless steel circular hollow section columns performed in Refs [5.15, 5.16].

The buckling analysis of cold-formed high strength stainless steel stiffened and unstiffened slender hollow section columns was highlighted by Ellobody [5.22] through numerical investigation. Nonlinear 3D finite element models were developed by the author to highlight the structural benefits of using stiffeners to strengthen slender square and rectangular hollow section columns. The construction material was high strength duplex stainless steel, which is austenitic-ferritic stainless steel that is approximately equivalent to EN 1.4462 and UNS S31803. The columns were compressed between fixed ends at different column lengths. The column strengths, load-shortening curves as well as failure modes were predicted for the stiffened and unstiffened slender hollow section columns. An extensive parametric study was conducted to study the effects of cross section geometries on the strength and behavior of the stiffened and unstiffened columns. The investigation has shown that the high strength stainless steel stiffened slender hollow section columns offer a considerable increase in the column strength over that of the unstiffened slender hollow section columns. The column strengths predicted from the parametric study were compared with the design strengths calculated using the American Specification [5.17], Australian/New Zealand Standard [5.18], and European Code [5.19] for cold-formed stainless steel structures. The study [5.22] has shown that the design strengths obtained using the three specifications are generally conservative for the cold-formed stainless steel unstiffened slender square and rectangular hollow section columns, but slightly unconservative for the stiffened slender square and rectangular hollow section columns. The 4-node doubly curved shell elements with reduced integration (S4R) were used to model the buckling behavior of the stiffened columns. The mesh size used in the model was approximately 20×10 mm (length by width). The load was applied in increments using the modified Riks method available in the ABAQUS [1.27] library. The nonlinear geometry parameter (*NLGEOM) was included to deal with the large displacement analysis. Both initial local and overall geometric imperfections were found in the column specimens.

Zhang et al. [5.23] have presented combined experimental and numerical investigations on cold-formed steel channels with inclined simple edge stiffeners compressed between pinned ends. The experimental investigation comprised a total of 36 channel specimens having different cross sections with different edge stiffeners. The initial geometric imperfections and material properties of the specimens were measured in [5.23]. The failure modes predicted included local buckling, distortional buckling, flexural buckling, and interaction of these buckling modes. The study has indicated that inclined angle and loading position significantly affect the ultimate load-carrying capacity and failure mode of the channels. The numerical investigation presented in Ref. [5.23] proposed a nonlinear finite element model, which was verified against the tests. Geometric and material nonlinearities were included in the model. The 4-node 3D quadrilateral shell element with six degrees of freedom at each node (S4), as detailed in Section 3.2, was used in the finite element analyses. Eigenvalue buckling analyses and nonlinear load–displacement analyses were conducted in the study [5.23]. By varying the size of the elements, the finite element mesh used in the model was studied. It was found that good simulation results could be obtained by using the element size of approximately 20×10 mm (length by width) for the lip and 20×16 mm for the flange and web.

Becque and Rasmussen [5.24] have detailed a finite element model studying the interaction of local and overall buckling in stainless steel columns. The model incorporated nonlinear stress–strain behavior, anisotropy, enhanced corner properties, and initial imperfections. The model was verified against tests on stainless steel lipped channels. The finite element model was further used in parametric studies, varying both the cross-sectional slenderness and the overall slenderness. Three stainless steel alloys were considered in the finite element analyses. The results were compared with the governing design rules of the Australian [5.18], North American [5.17], and European [5.19] standards for stainless steel structures. A 4-node shell element with reduced integration (S4R) was selected from the ABAQUS [1.27] element library to model the columns.

Gao et al. [5.25] have studied the load–carrying capacity of thin–walled box-section stub columns fabricated by high strength steel through experimental and numerical investigations. The columns investigated were axially loaded having different geometries. The column strengths obtained from the study [5.25] were compared with the design strengths predicted using the American Iron and Steel Institute (AISI) code [5.26]. The finite element analyses have used a general-purpose software ANSYS [5.3]. Parametric studies were performed to investigate the ultimate strength of the high strength steel stub columns. The authors have proposed a formula to predict the loading capacity of the high strength steel stub column based on the experimental and numerical investigations. Both material and geometric nonlinearities were adopted in the calculations. Initial local and overall geometric imperfections were included in the finite element model. The residual stresses were also incorporated in the finite element model.

Theofanous and Gardner [5.27] have detailed combined experimental and numerical investigations highlighting the compressive structural response of the lean duplex stainless steel columns. The authors have carried out a total of 8 stub column tests and 12 long column tests on lean duplex stainless steel square and rectangular hollow sections. The mechanical material properties, geometric properties, and assessment of local and global geometric imperfections were measured in the study [5.27]. Nonlinear finite element analyses and parametric studies were performed to generate results over a wide range of cross-sectional and member slenderness. The authors have used the experimental and numerical results to assess the applicability of the Eurocode 3: Part 1−4 [5.19] provisions regarding the Class 3 slenderness limit and effective width formula for internal elements in compression and the column buckling curve for hollow sections to lean duplex structural components. The authors have used the published test data to validate the finite element models. The general-purpose finite element software ABAQUS [1.27] was used for all numerical studies reported in the paper. The finite element simulations followed the proposals regarding numerical modeling of stainless steel components previously reported by one of the authors in Refs [5.28,5.29]. The measured geometric properties for stub columns and long columns have been employed in the finite element models. The 4-node doubly curved shell element with reduced integration S4R has been used in the study [5.27]. The authors have noted that the geometry, boundary conditions, applied loads, and failure modes of the tested components were symmetric; therefore, symmetry was exploited in the finite element modeling with suitable boundary conditions applied along the symmetry axes, enabling significant savings in computational time. In the stub columns [5.27], only a quarter of the section was modeled, whereas for the long columns, half of the cross section was discretized. For both stub columns and long columns, the full length of component was modeled. All degrees of freedom were restrained at the end cross sections of the stub column models, apart from the vertical translation at the loaded end, which was constrained via kinematic coupling to follow the same vertical displacement.

Goncalves and Camotim [5.30] have presented a geometrically and materially nonlinear generalized beam theory formulation. The finite element analysis presented aimed to evaluate nonlinear elastoplastic equilibrium paths of thin-walled metal bars and associated collapse loads. This finite element investigation was an extension to previously reported study by the authors [5.31] by including the geometrically nonlinear effects. The authors have assumed that the plate-like bending strains are assumed to be small, but the membrane strains are calculated exactly. The study used both stress-based and stress resultant-based generalized beam theory approaches in a 3-node beam finite element. The study has shown that the stress-based formulation is generally more accurate, but the stress resultant-based formulation makes it possible to avoid numeric integration in the through-thickness direction of the walls. The investigation [5.30] comprised several numerical examples. The results obtained from the study were also compared with that obtained using standard 2D solid and shell finite element analyses.

5.3. FINITE ELEMENT MODELING AND EXAMPLE 1

The first example presented in this chapter is for cold-formed high strength stainless steel columns, which were occasionally mentioned in the previous chapters for explanation. The column tests were carried out by Young and Lui [1.28,1.29] and provided the experimental ultimate loads and failure modes of columns compressed between fixed ends, as shown in Figure 3.1. The stainless steel columns had square and rectangular hollow sections having different geometries and lengths. The details of column specimens were found in [1.28,1.29] and no intention to repeat the published information in this book. However, it should be mentioned that the experimental program presented was well planned such that 22 tests were carefully conducted. The tests were also well instrumented such that the experimental results were used in the verification of the finite element models developed by Ellobody and Young [1.30]. The experimental program presented in [1.28,1.29] agrees well with the criteria previously discussed, in Section 1.3, for a successful experimental investigation. The cross section dimensions, material properties of the flat and corner portions of the specimens, initial local and overall geometric imperfections as well as residual stresses were measured as detailed in [1.28,1.29]. The results obtained from the tests [1.28,1.29] included the column strengths, load—axial shortening relationships and failure modes, which once again conforms to the criteria mentioned in Section 1.3. The results obtained have provided enough information for finite element models to be developed.

The tests reported by Young and Lui [1.28,1.29] were modeled by Ellobody and Young [1.30]. The general-purpose finite element software ABAQUS [1.27] was used to simulate the cold-formed high strength stainless steel columns. The authors have developed a nonlinear finite element model that accounted for the measured geometry, initial local and overall geometric imperfections, residual stresses, and nonlinear material properties. The authors have performed two types of analyses. The first analysis was eigenvalue buckling analysis, which is mentioned previously as linear elastic analysis performed using the (*BUCKLE) procedure available in the ABAQUS [1.27] library. The second analysis was the load—displacement geometrically and materially nonlinear analysis, which follows the eigenvalue buckling analysis. The ultimate loads, failure modes, and axial shortenings as well as any other required data were determined from the second analysis. The initial imperfections, residual stresses, and material nonlinearity were also included in the second analysis.

The 4-node doubly curved shell elements with reduced integration (S4R) were used in Ref. [1.30] to model the buckling behavior of cold-formed high strength stainless steel columns. In order to choose the finite element mesh that provides accurate results with minimum computational time, convergence studies were conducted by the authors. It was found that the mesh size of 20×10 mm (length by width) provides adequate accuracy and minimum computational time in modeling the flat portions of cold-formed high strength stainless steel columns, while finer mesh was used at the corners. The boundary conditions and load application were identical to the tests [1.28,1.29]. The load was applied in increments using the modified Riks method available in the ABAQUS [1.27] library. The nonlinear geometry parameter (*NLGEOM) was included to deal with the large displacement analysis. The measured stress—strain curves for flat and corner portions of the specimens [1.28,1.29] were used in the finite element analyses [1.30]. The material behavior provided by ABAQUS [1.27] allows for a multilinear stress—strain curve to be used, as described in Section 3.4. Cold-formed high strength stainless steel columns with large overall depth-to-plate thickness (D/t) ratio are likely to fail by local buckling or interaction of local and overall buckling depending on the column length and dimension. On the other hand, columns with small D/t ratio are likely to fail by yielding or overall buckling. Both initial local and overall geometric imperfections were found in the tested columns. Hence, superposition of local buckling mode as well as overall buckling mode with measured magnitudes was used in the finite element analyses. These buckling modes were obtained from the eigenvalue buckling analyses of the column with large D/t ratio as well as small D/t ratio to ensure local and overall buckling occurs, respectively. Only the lowest buckling mode (eigenmode 1) was used in the eigenvalue analyses.

The measured residual stresses [1.28,1.29] were included in the finite element model to ensure accurate modeling of the behavior of cold-formed high strength stainless steel columns. Measured residual stresses were implemented in the finite element model by using the ABAQUS (*INITIAL CONDITIONS, TYPE = STRESS) parameter (see details in Section). The material tests of flat and corner coupons considered the effect of bending residual stresses; hence, only the membrane residual stresses were modeled in this study. The magnitudes and distributions of the membrane residual stresses in the flat and corner portions of the columns were reported by Young and Lui [1.28,1.29]. A preliminary load step to allow equilibrium of the residual stresses was defined before the application of loading.

In the verification of the finite element model [1.30], a total of 22 cold-formed high strength stainless steel columns were analyzed. A comparison between the experimental and finite element analysis results was carried out. The main objective of this comparison is to verify and check the accuracy of the finite element model. The comparison of the ultimate test and finite element analysis loads (P_{Test} and P_{FE}), load—axial shortening relationships from the tests, and finite element analysis and failure modes obtained experimentally and numerically were compared in Ref. [1.30]. The comparison of the ultimate loads has shown that good agreement has been achieved between both results for most of the columns. A maximum difference of 8% was observed between experimental and numerical results. Three failure modes observed experimentally were verified by the finite element model. The failure modes were yielding failure (Y), local buckling (L), and flexural buckling (F). Figure 5.1 shows the deformed shape of rectangular hollow section column having a length of 3000 mm and the cross section dimensions shown in Figure 3.1, observed experimentally and numerically using the finite element model. The failure modes observed in the test and confirmed using the finite element analysis were interaction of local and flexural buckling (L + F). It can be seen that the finite element model accurately predicted the failure modes of the column observed in the test.

The study [1.30] has also investigated the effect of residual stresses on the behavior of cold-formed high strength stainless steel columns of duplex material. It has shown that the measured membrane residual

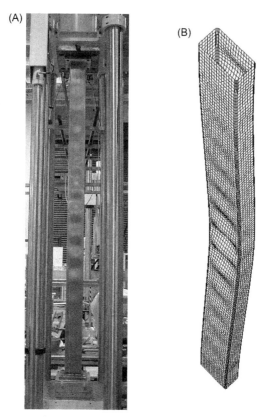

Figure 5.1 Comparison of experimental analysis (A) and finite element analysis (B) failure modes of high strength rectangular hollow section stainless steel column [1.30].

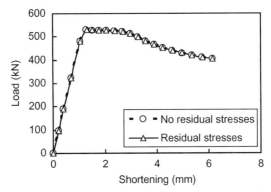

Figure 5.2 Load–axial shortening curves of high strength rectangular hollow section stainless steel column having a length of 600 mm [1.30].

stresses have a negligible effect on the ultimate load and load–shortening behavior. In Figure 5.2, the load versus axial shortening of the column having the cross section dimensions shown in Figure 3.1, and a length of 600 mm is presented. The curves were plotted with and without the simulation of the membrane residual stresses. It has shown that the ultimate load and behavior of the columns are almost identical. Therefore, in order to avoid the complexity of the analysis, the authors did not include the residual stresses in the parametric studies.

Following the verification of the finite element model, the authors [1.30] have performed parametric studies to study the effects of cross section geometries on the strength and behavior of the columns. A total of 42 columns were analyzed in the parametric study to generate more data outside the range covered by the experimental investigation [1.28,1.29]. The ultimate loads (P_{FE}) and failure modes were predicted from the parametric studies, which were considered as new information regarding the high strength stainless steel columns investigated. The results obtained from the parametric study can be used to extend the limits of design specified in the current codes of practice as conducted in the study [1.30]. The column strengths predicted from the parametric studies were compared with the unfactored design strengths calculated using the American [5.17], Australian/New Zealand [5.18], and European [5.19] specifications for cold-formed stainless steel structures. Table 5.1 shows an example of the comparison between the column strengths obtained from finite element analysis (P_{FE}) and design calculations, the nominal (unfactored) design strengths P_{ASCE} obtained using the American Specification [5.17],

Table 5.1 Comparison of Column Strengths Obtained from Finite Element Analysis and Design Specifications for High Strength Rectangular Hollow Section Stainless Steel Columns [1.30]

Specimen	P_{FE} (kN)	P_{ASCE} (kN)	$P_{AS/NZS}$ (kN)	P_{EC3} (kN)	$\dfrac{P_{FE}}{P_{ASCE}}$	$\dfrac{P_{FE}}{P_{AS/NZS}}$	$\dfrac{P_{FE}}{P_{EC3}}$
RT3L300	392.5	398.5	398.5	398.5	0.98	0.98	0.98
RT3L650	376.3	398.5	398.5	398.5	0.94	0.94	0.94
RT3L1000	355.8	397.4	398.5	390.2	0.90	0.89	0.91
RT3L1500	327.6	330.0	318.1	333.9	0.99	1.03	0.98
RT3L2000	299.7	278.4	264.0	268.0	1.08	1.14	1.12
RT3L2500	274.6	234.5	214.4	207.0	1.17	1.28	1.33
RT3L3000	217.3	193.8	170.7	158.4	1.12	1.27	1.37
Mean	–	–	–	–	1.03	1.08	1.09
COV	–	–	–	–	0.097	0.145	0.172

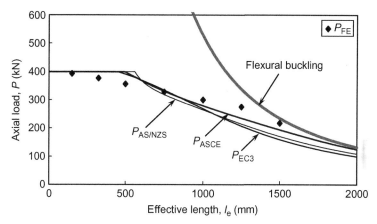

Figure 5.3 Comparison of column strengths obtained from finite element analysis and design specifications for high strength stainless steel columns [1.30].

$P_{AS/NZS}$ obtained using the Australian/New Zealand Standard [5.18], and P_{EC3} obtained using the European Code [5.19]. The example shown in Table 5.1 is a rectangular hollow section column, as shown in Figure 3.1, having an overall depth (D) of 90 mm, overall width (B) of 45 mm, and a plate thickness of 3 mm. The column strengths were also plotted on the vertical axis of Figure 5.3, as reported in Ref. [1.30], while the horizontal axis was plotted as the effective length (l_e) that is assumed equal to one-half of the column length for the fixed-ended columns.

5.4. FINITE ELEMENT MODELING AND EXAMPLE 2

The second example presented in this chapter is the aluminum alloy columns, which was carried out by Zhu and Young [5.32] and provided the experimental ultimate loads, load—axial shortening relationships and failure modes of the aluminum alloy columns. The authors have used the test results to develop a nonlinear finite element model simulating the buckling behavior of the columns as detailed in [5.12]. The tested columns [5.32] had square and rectangular hollow sections and were compressed between fixed ends, as shown in Figure 5.4. The test specimens were fabricated by extrusion using normal strength 6063-T5 and high strength 6061-T6 heat-treated aluminum alloys. The test program included 25 fixed-ended columns with both ends welded to aluminum end plates, and 11 fixed-ended columns without the welding of end plates. Therefore, the authors have used in Ref. [5.12] the term "welded column," which refers to a specimen with transverse welds at the ends of the column, whereas the term "nonwelded column" refers to a specimen without transverse welds. The testing condition of the welded and nonwelded columns is identical, other than the absence of welding in the nonwelded columns. The details of column specimens were found in Ref. [5.32] and, once again, no intention to repeat the published information in this book. Details should

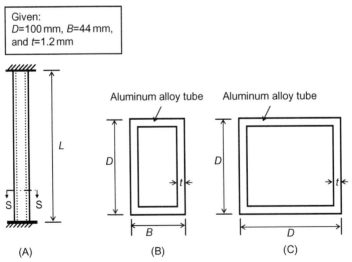

Figure 5.4 Example 2 of a fixed-ended aluminum alloy rectangular hollow section column [5.12,5.32]. (A) Fixed-ended rectangular hollow section column. (B) Rectangular hollow section (section S-S). (C) Square hollow section (section S-S).

be referred to Refs [5.12,5.32]. However, it should be mentioned that the experimental program presented was well planned such that 25 tests were accurately conducted. The tests were well instrumented such that the experimental results were used in the verification of the finite element models developed by Zhu and Young [5.12]. The nonwelded and welded material properties for each series of specimens were determined by longitudinal tensile coupon tests as detailed by Zhu and Young [5.32]. Initial overall geometric imperfections were measured for all specimens prior to testing. Initial local geometric imperfections were also measured for some specimens. The details of the measurements were found by Zhu and Young [5.32].

The general-purpose finite element software ABAQUS [1.27] was used in the analysis for the simulation of aluminum alloy fixed-ended columns tested by Zhu and Young [5.32]. The measured geometry, initial overall and local geometric imperfections, and material properties of the test specimens were used in the finite element model. The model was based on the center-line dimensions of the cross sections. Residual stresses were not included in the model. This is because in extruded aluminum alloy profiles, whatever be the heat treatment, residual stresses have very small values; for practical purpose, these have a negligible effect on load–bearing capacity [5.14]. The authors have also performed two types of analyses. The first analysis was eigenvalue buckling analysis (see Section 4.3) which is as previously mentioned a linear elastic analysis performed using the (*BUCKLE) procedure available in the ABAQUS [1.27] library. The second analysis was the load–displacement geometrically and materially nonlinear analysis, which follows the eigenvalue buckling analysis. The ultimate loads, failure modes, and axial shortenings as well as any other required data were determined from the second analysis. The initial imperfections and material nonlinearity were also included in the second analysis.

The 4-node doubly curved shell elements with reduced integration (S4R) were used in Ref. [5.12] to model the buckling behavior of the aluminum alloy columns. The size of the finite element mesh of 10×10 mm (length by width) was used in the modeling of the columns. The authors have mentioned that the heat-treated aluminum alloys suffer loss of strength in a localized region when welding is involved, and this is known as heat-affected zone (HAZ) softening. The welded columns were modeled by dividing the columns into different portions along the column length so that the HAZ softening at both ends of the welded columns was included in the simulation. The welded columns were separated into three parts, the HAZ regions at both ends of the columns, and the main body of the columns that are not affected by welding. Different mesh sizes were considered in the HAZ regions. The authors have noted that the American Specification [5.33] and Austrian/New Zealand Standard [5.34] for aluminum structures specified that the HAZ shall be taken as 1 in. (25.4 mm). However, the European Code [5.35] for aluminum structures assumed the HAZ extends to 30 mm for a TIG weld while the section thickness is less than 6 mm.

The boundary conditions and load application were identical to the tests [5.32]. The load was applied in increments using the modified Riks method, available in the ABAQUS [1.27] library. The nonlinear geometry parameter (*NLGEOM) was included to deal with the large displacement analysis. The measured stress–strain curves of the aluminum alloy specimens [5.32] were used in the finite element analyses [5.12]. The material behavior provided by ABAQUS [1.27] allows for a multilinear stress–strain curve to be used.

The developed nonlinear finite element model detailed in Ref. [5.12] was verified against the experimental results [5.32]. The authors have compared the ultimate loads obtained numerically (P_{FEA}) with that obtained experimentally (P_{Exp}), and good agreement was achieved. The mean value of P_{Exp}/P_{FEA} ratio was 1.02 with the corresponding coefficient of variation (COV) of 0.045 for the nonwelded columns, as discussed in [5.12]. For the welded columns, both the ultimate loads predicted by the finite element analysis using the HAZ extension of 25 mm (P_{FEA25}) and 30 mm (P_{FEA30}) were compared with the experimental results. The authors have shown that the P_{FEA25} are in better agreement with the experimental ultimate loads compared with the P_{FEA30}. The observed failure modes obtained experimentally [5.32] and confirmed numerically [5.12] included local buckling (L), flexural buckling (F), interaction of local and flexural buckling (L + F), and failure in the HAZ. Figure 5.5 shows the comparison of the failure modes obtained from the test and predicted by the finite element analysis, as previously published in [5.12], for the high strength aluminum alloy nonwelded rectangular hollow section specimen having an overall depth (D) of 100, overall width (B) of 44, plate thickness (t) of 1.2, and a length of 1000 mm. Figure 5.6 shows the load-shortening curves obtained experimentally and numerically for the high strength aluminum alloy welded rectangular hollow section specimen having a D of 100, B of 44, t of 1.2, and a length of 2350 mm. The load-shortening curves predicted by the finite element analysis using the HAZ extension of 25 and 30 mm are

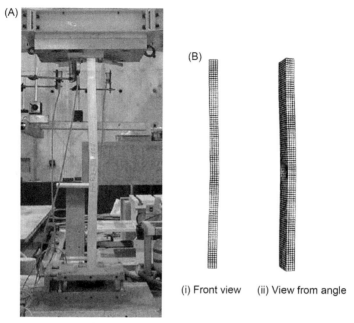

(i) Front view   (ii) View from angle

Figure 5.5 Comparison of failure modes obtained from experimental analysis (A) and finite element analysis (B) of rectangular hollow section aluminum alloy column having a length of 1000 mm [5.12].

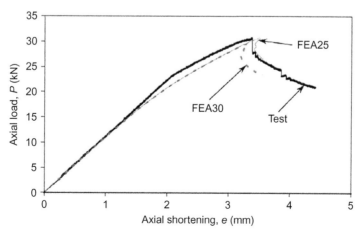

Figure 5.6 Load–axial shortening curves obtained experimentally and numerically for rectangular hollow section aluminum alloy column having a length of 2350 mm [5.12].

shown in Figure 5.6. The authors have shown that good agreement was found between the experimental and numerical curves, which demonstrated the reliability of the finite element analysis predictions.

5.5. FINITE ELEMENT MODELING AND EXAMPLE 3

The third example presented in this chapter is the cold-formed steel plain angle columns (Figure 5.7), which Young tested [5.36] and provided the experimental ultimate loads, load–axial shortening relationships, and failure modes of the columns. The test program included 24 fixed-ended cold-formed steel plain angle columns. The test program agrees well with the criteria previously discussed, in Section 1.3, for a successful experimental investigation. The authors have used the test results to develop a nonlinear finite element model simulating the buckling behavior of the columns as detailed by Ellobody and Young [4.9]. The test specimens were brake-pressed from high strength zinc-coated grades G500 and G450 structural steel sheets having nominal yield stresses of 500 and 450 MPa, respectively, and conformed to the Australian Standard AS 1397 [5.37]. Each specimen was cut to a specified length of 250, 1000, 1500, 2000, 2500, 3000, and 3500 mm. Three series of plain angles were tested, having a nominal flange width of 70 mm. The nominal plate thicknesses were 1.2, 1.5, and 1.9 mm. The three series were labeled

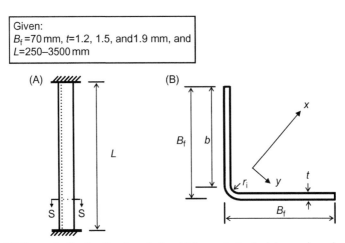

Figure 5.7 Example 3 of a fixed-ended cold-formed steel plain angle column [4.9, 5.36]. (A) Fixed-ended plain angle section column. (B) Plain angle section (Section S-S).

P1.2, P1.5, and P1.9 according to their nominal thickness. The measured inside corner radius was 2.6 mm for all specimens. The measured cross section dimensions of the test specimens are detailed by Young [5.36]. The measured flat flange width-to-thickness ratio was 57.9, 45.0, and 35.8 for Series P1.2, P1.5, and P1.9, respectively. The test specimens are labeled such that the test series and specimen length could be identified from the label. For example, the label "P1.2L1000" defines the specimen belonged to test Series P1.2, and the fourth letter "L" indicates the length of the specimen followed by the nominal column length of the specimen in millimeters (1000 mm).

The material properties of the flange (flat portion) of the specimens for each series were determined by tensile coupon tests. The coupons were taken from the center of the flange plate in the longitudinal direction of the finished specimens. The coupon dimensions and the tests conformed to the Australian Standard AS 1391 [5.38] for the tensile testing of metals using 12.5 mm wide coupons of gauge length 50 mm. The Young's modulus (E), the measured static 0.2% proof stress ($\sigma_{0.2}$), the measured elongation after fracture (ε) based on a gauge length of 50 mm, and the tensile coupon tests of the flat portions were measured as detailed by Young [5.36]. The initial overall geometric imperfections of the specimens were measured prior to testing. The maximum overall imperfections at mid-length were 1/2950, 1/2150, and 1/1970 of the specimen length for Series P1.2, P1.5, and P1.9, respectively. The measured overall geometric imperfections of each test specimen are detailed by Young [5.36]. A servo-controlled hydraulic testing machine was used to apply compressive axial force to the specimen. The fixed-ended bearings were designed to restrain against the minor and major axis rotations as well as twist rotations and warping. Displacement control was used to allow the tests to be continued in the post-ultimate range. The column tests are detailed by Young [5.36]. The initial local geometric imperfections, residual stresses, and corner material properties of the tested plain angle specimens were not reported by Young [5.36]. However, the values of these measurements are important for finite element analysis. Hence, the initial local imperfections, residual stresses, and corner material properties of the angle specimens belonging to the same batched as the column test specimens were measured and reported by Ellobody and Young [4.9].

The finite element program ABAQUS [1.27] was used in the analysis of plain angle columns tested by Young [5.36]. The model used the measured geometry, initial local and overall geometric imperfections, residual stresses, and material properties as detailed by Ellobody and Young [4.9]. Since buckling of plain angle columns is very sensitive to large strains, the S4R element was used in this study to ensure the accuracy of the results. In order to choose the finite element mesh that provides accurate results with minimum computational time, convergence studies were conducted. It is found that 10×10 mm (length by width) ratio provides adequate accuracy in modeling the flat portions of plain angles while finer mesh was used at the corner. Following the testing procedures for Series P1.2, P1.5, and P1.9, the ends of the columns were fixed against all degrees of freedom except for transitional displacement at the loaded end in the direction of the applied load. The nodes other than the two ends were free to translate and rotate in any directions. The load was applied in increments using the modified Riks method available in the ABAQUS [1.27] library. The load was applied as static uniform loads at each node of the loaded end which is identical to the experimental investigation. The nonlinear geometry parameter (NLGEOM) was included to deal with the large displacement analysis. The measured stress–strain curves for flat portions of Series P1.2, P1.5, and P1.9 were used in the analysis. The material behavior provided by ABAQUS [1.27] allows for a multi-linear stress–strain curve to be used, as described in Section 3.4.

Cold-formed steel plain angle columns with very high b/t ratio are likely to fail by pure local buckling. On the other hand, columns with very low b/t ratio are likely to fail by overall buckling. Both initial local and overall geometric imperfections are found in columns as a result of the fabrication process. Hence, superposition of local buckling mode as well as overall buckling mode with measured magnitudes is recommended for accurate finite element analysis. These buckling modes can be obtained by carrying eigenvalue analysis of the column with very high b/t ratio and very low b/t ratio to ensure local and overall buckling, occurs respectively. The shape of a local buckling mode as well as overall buckling mode is found to be the lowest buckling mode (eigenmode 1) in the analysis. This technique is used in this study to model the initial local and overall imperfections of the columns. Stub columns having very short length can be modeled for local imperfection only. Since all buckling modes predicted by ABAQUS [1.27] eigenvalue analysis are generalized to 1.0, the buckling modes are factored by the measured magnitudes of the initial local and overall geometric imperfections. Figure 5.8 shows the unfactored local and overall imperfection buckling modes for angle specimen P1.9L1500. More details regarding modeling of geometric

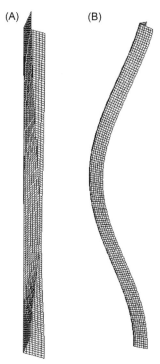

Figure 5.8 Initial geometric imperfection modes (eigenmode 1) for plain angle column P1.9L1500 [4.9]. (A) Local imperfection. (B) Overall imperfection.

imperfections can be found in Section 4.3. Measured residual stresses are implemented in the finite element model by using the ABAQUS [1.27] (*ININTIAL CONDITIONS, TYPE = STRESS) parameter. The flat and corner coupons material tests took into consideration of the bending residual stresses effect, hence, only the membrane residual stresses were modeled. Detailed information regarding modeling residual stresses can be found in Section 3.6.

In the verification of the finite element model, a total of 21 cold-formed steel plain angle columns were analyzed. A comparison between the experimental results and the results of the finite element model was carried out. The comparison of the ultimate loads (P_{Test} and P_{FE}), axial shortening (e_{Test} and e_{FE}) at the ultimate loads, and failure modes obtained experimentally and numerically are given in Table 5.2. Figure 5.9 plotted the relationship between the ultimate load and the column effective length ($l_{ey} = L/2$) for Series P1.2, P1.5, and P1.9, where L is the actual column length. The column curves show the experimental ultimate loads

Table 5.2 Comparison between Test and FE Results for Cold-Formed Steel Plain Angle Columns [4.9]

Specimen	Test			FE			Test/FE	
	P_{Test} (kN)	e_{Test} (mm)	Failure Mode	P_{FE} (kN)	e_{FE} (mm)	Failure Mode	$\dfrac{P_{Test}}{P_{FE}}$	$\dfrac{e_{Test}}{e_{FE}}$
P1.2L250	23.8	0.54	L	24.9	0.61	L	0.96	0.89
P1.2L1000	18.7	1.10	F + FT	18.1	0.96	F + FT	1.03	1.14
P1.2L1500	15.2	0.82	F + FT	15.7	0.70	F + FT	0.97	1.17
P1.2L2000	12.6	1.66	F + FT	11.7	1.74	F + FT	1.08	0.95
P1.2L2500	11.6	1.25	F + FT	10.1	1.04	F + FT	1.15	1.20
P1.2L3000	8.0	1.07	F + FT	8.6	0.98	F	0.93	1.15
P1.2L3500	5.8	1.03	F + FT	6.4	0.89	F	0.91	1.16
Mean	–	–	–	–	–	–	1.00	1.09
COV	–	–	–	–	–	–	0.086	0.111
P1.5L250	39.6	0.70	L	37.8	0.83	L	1.05	0.84
P1.5L1000	31.0	1.20	F + FT	31.5	1.39	F + FT	0.98	0.86
P1.5L1500	25.2	1.25	F + FT	25.5	1.03	F + FT	0.99	1.21
P1.5L2000	17.5	1.27	F + FT	19.7	1.07	F + FT	0.89	1.19
P1.5L2500	15.7	1.42	F + FT	16.0	1.19	F + FT	0.98	1.19
P1.5L3000	13.1	1.32	F + FT	14.2	1.21	F	0.92	1.09
P1.5L3500	11.5	1.36	F + FT	12.5	1.29	F	0.92	1.05
Mean	–	–	–	–	–	–	0.96	1.06
COV	–	–	–	–	–	–	0.057	0.147
P1.9L250	57.7	0.80	L	60.6	0.95	L	0.95	0.84
P1.9L1000	47.8	1.40	FT	49.1	1.55	F + FT	0.97	0.90
P1.9L1500	35.6	1.45	F + FT	38.4	1.27	F + FT	0.93	1.14
P1.9L2000	27.1	1.66	F + FT	30.8	1.45	F + FT	0.88	1.14
P1.9L2500	22.4	1.88	F + FT	24.3	1.62	F + FT	0.92	1.16
P1.9L3000	14.8	1.22	F + FT	16.7	1.14	F	0.89	1.07
P1.9L3500	14.4	1.15	F + FT	15.4	1.22	F	0.94	0.94
Mean	–	–	–	–	–	–	0.93	1.03
COV	–	–	–	–	–	–	0.035	0.128

together with that obtained by the finite element method. It can be seen that good agreement has been achieved between both results for most of the columns. The finite element results are slightly higher than that of the test strengths for Series P1.5 and P1.9. A maximum difference of 15% was observed between experimental and numerical results for P1.2L2500 column. The mean values of P_{Test}/P_{FE} ratio are 1.00, 0.96, and 0.93 with the corresponding COV of 0.086, 0.057, and 0.035 for Series P1.2, P1.5, and P1.9, respectively, as given in Table 5.2. The mean values of e_{Test}/e_{FE} ratio are 1.09, 1.06, and 1.03 with the COV of 0.111, 0.147, and 0.128

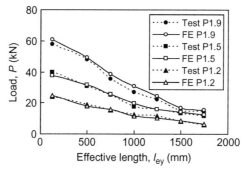

Figure 5.9 Ultimate loads obtained experimentally and numerically for plain angle columns for Series P1.2, P1.5, and P1.9 [4.9].

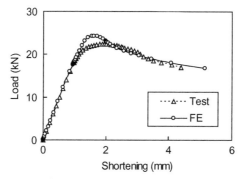

Figure 5.10 Comparison of load–axial shortening curves obtained experimentally and numerically for column P1.9L2500 [4.9].

for Series P1.2, P1.5, and P1.9, respectively. Generally, good agreement has been achieved for most of the columns. Three modes of failure have been observed experimentally and verified by the finite element model. The failure modes are the local buckling (L), flexural buckling (F), and flexural-torsional buckling (FT).

Figure 5.10 shows the ultimate load against the axial shortening behavior of column P1.9L2500 that has a length of 2500 mm. The curve has been predicted by the finite element model and compared with the test curve. It can be shown that both the column stiffness and behavior reflect good agreement between experimental and finite element results. The failure modes observed in the test of P1.9L2500 were interaction of flexural and flexural-torsional buckling (F + FT). The same failure mode has been confirmed numerically by the model shown in Figure 5.11.

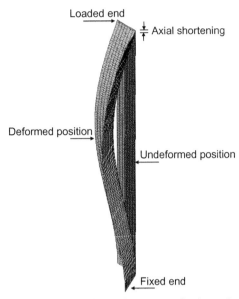

Figure 5.11 Failure mode and undeformed position of column P1.9L2500 [4.9].

It can be noticed that the cross section of the column was shifted and twisted from its undeformed position.

The main design rules used in the study [4.9] are specified in the AISI [5.39] and the AS/NZS [5.40]. Using the design rules for concentrically loaded compression members in the AISI Specification [5.39] and AS/NZS [5.40], the nominal axial strength (P_n) were calculated as follows:

$$P_n = A_e F_n \tag{5.1}$$

where A_e is the effective area and F_n is the critical buckling stress. The critical buckling stress (F_n) is calculated as follows:

$$F_n = (0.658^{\lambda_c^2})F_y \quad \text{for} \quad \lambda_c \leq 1.5 \tag{5.2}$$

$$F_n = \left[\frac{0.877}{\lambda_c^2}\right]F_y \quad \text{for} \quad \lambda_c > 1.5 \tag{5.3}$$

where λ_c is the nondimensional slenderness calculated as follows:

$$\lambda_c = \sqrt{\frac{F_y}{F_e}} \tag{5.4}$$

where F_y is the yield stress which is equal to the 0.2% proof stress $(\sigma_{0.2})$ and F_e is the least of the elastic flexural, torsional, and flexural-torsional buckling stress determined in accordance with Sections C4.1−C4.3 of the AISI Specification and Sections 3.4.1−3.4.4 of the AS/NZS.

Young [5.36] concluded that the design strengths obtained using the AISI Specification and AS/NZS for the tested cold-formed steel plain angle columns are generally quite conservative. Hence, Eqs (5.2) and (5.3) have been modified as follows:

$$F_n = (0.5^{\lambda_c^2})F_y \quad \text{for} \quad \lambda_c \leq 1.4 \qquad (5.5)$$

$$F_n = \left[\frac{0.5}{\lambda_c^2}\right]F_y \quad \text{for} \quad \lambda_c > 1.4 \qquad (5.6)$$

where the slenderness (λ_c) is calculated as that in Eq. (5.4) with the exception that the elastic buckling stress (F_e) is determined from the flexural buckling only in accordance with Section C4.1 of the AISI Specification and Section 3.4.2 of the AS/NZS. The column design strength (P_P) proposed by Young [5.36] was then computed as $P_P = A_e F_n$. It has been shown that the design strengths obtained using the proposed Eqs (5.5) and (5.6) compared well with the test strengths conducted by Young [5.36].

After the verification of the finite element model, parametric studies were carried out to study the effects of cross section geometries on the strength and behavior of angle columns. A total of 35 plain angle columns were analyzed in the parametric study. Five series of column P0.8, P1.0, P2.6, P4.2, and P10.0 having plate thickness of 0.8, 1.0, 2.6, 4.2, and 10.0 mm, respectively were studied. All angle sections had the overall flange width of 70 mm which is having the same flange width as the test specimens. The five series had the flat flange width-to-thickness ratio (b/t) of 85, 65, 25, 15, and 5 for Series P0.8, P1.0, P2.6, P4.2, and P10.0, respectively. Each series of columns consists of seven column lengths of 250, 1000, 1500, 2000, 2500, 3000, and 3500 mm. The maximum initial local geometric imperfection magnitude was taken as the measured value of 0.14% of the plate thickness. The maximum initial overall geometric imperfection magnitude was taken as the average of the measured maximum overall imperfections of the tested series which is equal to $L/2360$, where L is the column length.

The results of the parametric study were compared with the nominal (unfactored) design strengths obtained using the AISI Specification and AS/NZS as well as compared with the design strengths obtained using the equations proposed by Young [5.36]. Figures 5.12−5.16 show a comparison between the finite element results (P_{FE}) with the nominal (unfactored) design strengths (P_n) obtained using the AISI Specification and AS/NZS, the design strengths (P_F) that consider flexural buckling only when calculating the elastic buckling stress (F_e), and the design strengths (P_P) proposed by Young [5.36]. The column curves are nondimensionalized with respect to the nominal stub column design strength (section

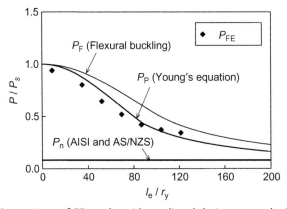

Figure 5.12 Comparison of FE results with predicted design strengths for Series P0.8 ($b/t = 85$) [4.9].

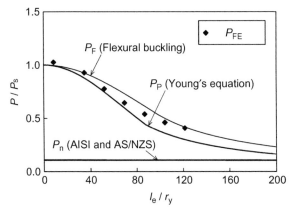

Figure 5.13 Comparison of FE results with predicted design strengths for Series P1.0 ($b/t = 65$) [4.9].

capacity) P_s, i.e., $P_s = A_e F_y$, where A_e is the effective area at yield stress of the flat portion (F_y) as shown on the vertical axis of Figures 5.12–5.16. The horizontal axis is plotted as l_e/r_y, where l_e is the effective length that assumed equal to one-half of the column length and r_y is the radius of gyration about the minor principal y-axis. It can be seen that the AISI and AS/NZS design strengths (P_n) are generally quite conservative for angle columns with b/t ratios of 85, 65, and 25 as shown in Figures 5.12–5.14. On the other hand, the AISI and AS/NZS design strengths overestimated the column strengths for most of the columns that have b/t ratios of 15 and 5, as shown in Figures 5.15 and 5.16. The

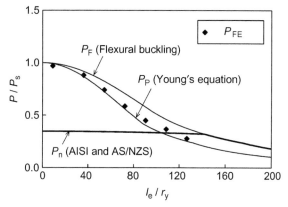

Figure 5.14 Comparison of FE results with predicted design strengths for Series P2.6 ($b/t = 25$) [4.9].

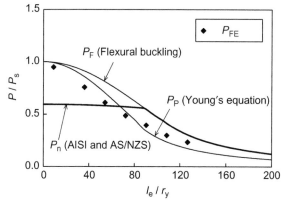

Figure 5.15 Comparison of FE results with predicted design strengths for Series P4.2 ($b/t = 15$) [4.9].

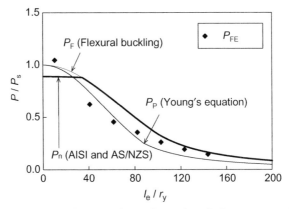

Figure 5.16 Comparison of FE results with predicted design strengths for Series P10.0 ($b/t = 5$) [4.9].

design strengths (P_F) that consider flexural buckling only overestimated the column strengths of all b/t ratios. Good agreement has been achieved between FE results and the design strengths (P_P) obtained using the equation proposed by Young [5.36] for most of the columns.

5.6. FINITE ELEMENT MODELING AND EXAMPLE 4

The fourth example presented in this chapter is the cold-formed stainless steel circular hollow section columns (Figure 5.17), which were tested by Young and Hartono [5.20], and they provided the experimental ultimate loads, load—axial shortening relationships, and failure modes of the columns. The test program included 16 fixed-ended cold-formed stainless steel circular hollow section columns. The test program agrees well with the criteria discussed in Section 1.3 for a successful experimental investigation. The authors have used the test results to develop a nonlinear finite element model simulating the buckling behavior of the columns as detailed by Young and Ellobody [5.15]. Three series (Series C1, C2, and C3) of circular hollow section columns were tested. The test specimens were cold-rolled from annealed flat strips of type 304 stainless steel. Each specimen was cut to a specified length (L) ranging from 550 to 3000 mm. The measured cross section dimensions of the test specimens are detailed by Young and Hartono [5.20]. The Series C1, C2, and C3 had an average measured outer diameter (D) of 89.0, 168.7, and 322.8 mm and an average thickness (t) of 2.78, 3.34, and 4.32 mm, respectively. The average measured outer

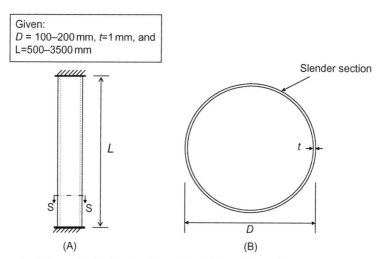

Given:
D = 100–200 mm, t=1 mm, and
L=500–3500 mm

Slender section

L

t

S S

D

(A)   (B)

Figure 5.17 Example 4 of a fixed-ended cold-formed stainless steel circular hollow section column [5.15,5.20]. (A) Fixed-ended slender circular hollow section. (B) Slender circular hollow section (section S-S).

diameter-to-thickness (D/t) ratio is 32.0, 50.5, and 74.7 for Series C1, C2, and C3, respectively. The test specimens are labeled such that the test series and specimen length could be identified from the label. For example, the label "C1L1000" defines the specimen belonged to test Series C1, and the letter "L" indicates the length of the specimen followed by the nominal column length of the specimen in millimeters (1000 mm).

The material properties of each series of specimens were determined by tensile coupon tests. The coupons were taken from the untested specimens at 90° from the weld in the longitudinal direction. The coupon dimensions and the tests conformed to the Australian Standard AS 1391 [5.38] for the tensile testing of metals using 12.5 mm wide coupons of gauge length 50 mm. The Young's modulus (E_0), the measured static 0.2% proof stress ($\sigma_{0.2}$), the measured elongation after fracture based on a gauge length of 50 mm were measured as detailed in Ref. [5.20]. The initial overall geometric imperfections of the specimens were measured prior to testing. The average values of overall imperfections at mid-length were 1/1715, 1/3778, and 1/3834 of the specimen length for Series C1, C2, and C3, respectively. The measured overall geometric imperfections for each test specimen are detailed by Young and Hartono [5.20]. The initial local geometric imperfections of the tested cold-formed stainless steel circular hollow section columns were not reported by Young and Hartono [5.20].

However, the values of the initial geometric imperfections are important for finite element analysis. Hence, the initial local geometric imperfections of the stainless steel circular hollow section specimen belonging to the same batch as the column test specimens are measured in this study and reported by Young and Ellobody [5.15]. A cold-formed stainless steel circular hollow section test specimen of 250 mm in length of Series C1 was used for the measurement of local imperfections. The maximum magnitude of local plate imperfection was 0.089 mm, which is equal to 3.2% of the plate thickness of the specimen belonged to Series C1. The same factor was used to predict the initial local geometric imperfections for Series C2 and C3.

The finite element program ABAQUS [1.27] was used to investigate the buckling behavior of the cold-formed stainless steel slender circular hollow section columns. The tests conducted by Young and Hartono [5.20] were modeled using the measured geometry, initial local and overall geometric imperfections, and material properties. In order to choose the finite element mesh that provides accurate results with minimum computational time, convergence studies were conducted. It is found that the mesh size around 10×10 mm (length by width) provides adequate accuracy and minimum computational time in modeling the cold-formed stainless steel circular hollow section columns. Following the testing procedures for Series C1, C2, and C3, the ends of the columns were fixed against all degrees of freedom except for the displacement at the loaded end in the direction of the applied load. The nodes other than the two ends were free to translate and rotate in any directions. The load was applied in increments using the modified Riks method available in the ABAQUS library [1.27]. The load was applied as static uniform loads at each node of the loaded end using displacement control which is identical to the experimental investigation. The nonlinear geometry parameter (*NLGEOM) was included to deal with the large displacement analysis. The measured stress–strain curves of Series C1, C2, and C3 were used in the analysis. The material behavior provided by ABAQUS [1.27] allows for a multilinear stress–strain curve to be used, as described in Section 3.4.

Cold-formed stainless steel columns with large D/t ratio are likely to fail by local buckling or interaction of local and overall buckling depending on the column length and dimension. Both initial local and overall geometric imperfections were found in the tested columns. Hence, superposition of local buckling mode as well as overall buckling mode with measured magnitudes is recommended in the finite element analysis.

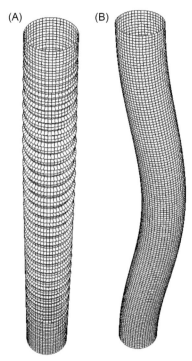

Figure 5.18 Initial geometric imperfection modes (eigenmode 1) for stainless cold-formed steel circular hollow section specimen C2L1500 [5.15]. (A) Local imperfection. (B) Overall imperfection.

These buckling modes can be obtained by carrying eigenvalue analyses of the column with large D/t ratio as well as small D/t ratio to ensure local and overall buckling occurs, respectively. Only the lowest buckling mode (eigenmode 1) was used in the eigenvalue analyses. This technique was used in this study to model the initial local and overall imperfections of the columns. Since all buckling modes predicted by ABAQUS [1.27] eigenvalue analysis are normalized to 1.0, the buckling modes were factored by the measured magnitudes of the initial local and overall geometric imperfections. Figure 5.18 shows the local and overall imperfection buckling modes for specimen C2L1500. More details regarding modeling of geometric imperfections can be found in Section 4.3.

The cold-formed stainless steel circular hollow section columns tested by Young and Hartono [5.20] were used to verify the finite element model. The comparison of the ultimate loads (P_{Test} and P_{FE}) and axial shortening (e_{Test} and e_{FE}) at the ultimate loads obtained experimentally

Table 5.3 Comparison between Test and Finite Element Results for Cold-Formed Stainless Steel Circular Hollow Section Columns [5.15,5.20]

	Test		FE			Test/FE	
Specimen	P_{Test} (kN)	e_{Test} (mm)	P_{FE} (kN)	e_{FE} (mm)	Failure Mode	$\dfrac{P_{Test}}{P_{FE}}$	$\dfrac{e_{Test}}{e_{FE}}$
C1L550	235.2	16.88	240.5	15.41	Y	0.98	1.10
C1L1000	198.4	10.26	206.8	10.89	Y	0.96	0.94
C1L1500	177.4	5.77	181.8	6.54	F	0.98	0.88
C1L2000	165.1	4.83	167.9	5.54	F	0.98	0.87
C1L2500	151.6	5.39	148.9	5.93	F	1.02	0.91
C1L3000	133.4	4.99	134.5	5.41	F	0.99	0.92
C2L550	495.6	9.41	522.0	8.32	Y	0.95	1.13
C2L1000	474.9	14.64	486.7	13.03	L	0.98	1.12
C2L1500	461.0	15.92	468.9	15.25	L + F	0.98	1.04
C2L2000	431.6	13.32	443.7	15.11	L + F	0.97	0.88
C3L1000	1123.9	8.05	1140.0	7.93	Y	0.99	1.02
C3L1500	1119.7	14.38	1130.0	13.12	Y	0.99	1.10
C3L2000	1087.8	14.53	1100.0	14.90	L	0.99	0.98
C3L2500	1045.7	19.12	1070.0	18.05	L	0.98	1.06
C3L3000	1009.5	15.64	1040.0	16.74	L	0.97	0.93
Mean	—		—		—	0.98	0.99
COV	—		—		—	0.016	0.095

and numerically are given in Table 5.3. It can be seen that good agreement has been achieved between both results for most of the columns. The mean value of P_{Test}/P_{FE} ratio is 0.98 with the corresponding COV of 0.016, as given in Table 5.3. The mean value of e_{Test}/e_{FE} ratio is 0.99 with the COV of 0.095. Three modes of failure have been observed experimentally and confirmed numerically by the finite element analysis. The failure modes were yielding failure (Y), local buckling (L), and flexural buckling (F). Figure 5.18 shows the applied load against the axial shortening behavior of column specimen C3L1000 that has an outer diameter of 322.8 mm and a length of 1000 mm. The curve has been predicted using the finite element analysis and compared with the test curve. There has been generally good agreement between experimental and finite element results. Figure 5.19 shows the deformed shape of column specimen C2L2000 observed experimentally and numerically using the FE analysis. The column has an outer diameter of 168.7 mm and a length of 2000 mm. The failure modes observed in the test were interaction of

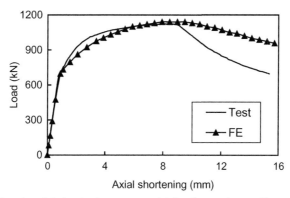

Figure 5.19 Load—axial shortening curves obtained experimentally and numerically for column C3L1000 [5.15].

local and flexural buckling (L + F). It can be seen that the finite element model accurately predicted the failure modes observed in the test.

After the verification of the finite element model developed in Ref. [5.15], parametric studies were performed to study the effect of local buckling on the strength and behavior of the slender circular hollow section columns. A total of 42 columns were analyzed in the parametric study. The columns are labeled such that the outer diameter and column length could be identified from the label. For example, the label "C100L1000" defines the circular hollow section column using a letter "C" followed by the value of the outer diameter in millimeters (100 mm) and the letter "L" indicates the length of the column followed by the column length in millimeters (1000 mm). Six series of slender circular hollow sections (Series C100, C120, C140, C160, C180, and C200) having the outer diameter of 100, 120, 140, 160, 180, and 200 mm, respectively, and a plate thickness of 1.0 mm were studied. The Series C100, C120, C140, C160, C180, and C200 had the outer diameter-to-thickness ratio (D/t) of 100, 120, 140, 160, 180, and 200, respectively. Each series of columns consists of seven column lengths of 500, 1000, 1500, 2000, 2500, 3000, and 3500 mm. The maximum initial local geometric imperfection magnitude was taken as the measured value of Series C1 which is equal to 3.2% of the plate thickness. The initial overall geometric imperfection magnitude was taken as the average of the measured overall imperfections of the Series C1 which is equal to $L/1715$, where L is the column length. The measured stress—strain curve of Series C1 was used in the parametric study.

The design rules specified in the ASCE [5.17] are based on the Euler column strength that requires the calculation of tangent modulus (E_t) using an iterative design procedure. The design rules specified in the EC3 [5.19] are based on the Perry curve that needs only the initial Young's modulus (E_0) and a number of parameters to calculate the design stress. The design rules specified in the AS/NZS [5.18] adopt either the Euler column strength or alternatively the Perry curve; the latter is used in this paper. The fixed ended columns were designed as concentrically loaded compression members and the effective length (l_e) were taken as one-half of the column length ($l_e = L/2$) as recommended by Young and Rasmussen [5.41]. In the three specifications, the effective area (A_e) is to account for local buckling of slender sections.

5.6.1 American Specification

The nominal (unfactored) design strength for concentrically loaded cylindrical tubular compression members in the ASCE [5.17] is calculated as follows:

$$P_{ASCE} = F_n A_e \tag{5.7}$$

The flexural buckling stress (F_n) that account for overall buckling is calculated as follows:

$$F_n = \frac{\pi^2 E_t}{(l_e/r)^2} \leq F_y \tag{5.8}$$

where E_t is the tangent modulus determined using Eq. (B-2) of the ASCE, l_e is the effectively length, r is the radius of gyration of the full cross section, and F_y is the yield stress that is equal to the static 0.2% proof stress ($\sigma_{0.2}$).

The effective area (A_e) that accounts for local buckling is calculated as follows:

$$A_e = \left[1 - \left(1 - \left(\frac{E_t}{E_0} \right)^2 \right) \left(1 - \frac{A_0}{A} \right) \right] A \tag{5.9}$$

where E_0 is the initial Young's modulus, A is the full cross-sectional area, and A_0 is the reduced cross-sectional area which is determined as follows:

$$A_0 = K_c A \leq A \quad \text{for} \quad \frac{D}{t} \leq \frac{0.881 E_0}{F_y} \tag{5.10}$$

and

$$K_c = \frac{(1 - C)(E_0/F_y)}{(8.93 - \lambda_c)(D/t)} + \frac{5.882C}{8.93 - \lambda_c} \tag{5.11}$$

where C is the ratio of effective proportional limit-to-yield strength as given in Table A17 of the ASCE [5.17], $\lambda_c = 3.084C$ with a limiting value of $(E_0/F_y)/(D/t)$, D is the outer diameter, and t is the plate thickness of the stainless steel tube.

5.6.2 Australian New/Zealand Standard

The unfactored design strength for concentrically loaded cylindrical tubular compression members in the AS/NZS [5.18] is calculated using the Perry curve as follows:

$$P_{AS/NZS} = F_n A_e \tag{5.12}$$

where

$$F_n = \frac{F_y}{\varphi + \sqrt{\varphi^2 - \lambda^2}} \leq F_y \tag{5.13}$$

$$\varphi = 0.5(1 + \eta + \lambda^2) \tag{5.14}$$

$$\eta = \alpha\left((\lambda - \lambda_1)^\beta - \lambda_0\right) \geq 0 \tag{5.15}$$

$$\lambda = \frac{l_e}{r}\sqrt{\frac{F_y}{\pi^2 E_0}} \tag{5.16}$$

The parameters α, β, λ_0, and λ_1 required for the calculation of the AS/NZS design strengths were calculated from the equations proposed by Rasmussen and Rondal [5.42]. The columns investigated in the parametric study had slenderness (λ) ranged from 0.043 to 0.06 calculated using Eq. (5.16), which covered the short to intermediate column slenderness. Hence, the results of the present study are limited to cold-formed stainless steel slender circular hollow sections for short to intermediate column slenderness. The effective area (A_e) is calculated in the same way as the ASCE [5.17], except for the reduction factor K_c as given in Eq. (5.11). In the AS/NZS [5.18], the reduction factor K_c is calculated as follows:

$$K_c = \frac{(1 - C)(E_0/F_y)}{(3.226 - \lambda_c)(D/t)} + \frac{0.178C}{3.226 - \lambda_c} \tag{5.17}$$

5.6.3 European Code

The unfactored design strength for concentrically loaded cylindrical tubular compression members in the EC3 [5.19] is calculated as follows:

$$P_{EC3} = \chi F_y A_e \tag{5.18}$$

where

$$\chi = \frac{1}{\varphi + \sqrt{\varphi^2 - \overline{\lambda}^2}} \leq 1 \tag{5.19}$$

$$\varphi = 0.5(1 + \alpha(\overline{\lambda} - \overline{\lambda}_0) + \overline{\lambda}^2) \tag{5.20}$$

$$\overline{\lambda} = \frac{l_e}{r} \sqrt{\frac{F_y \beta_A}{\pi^2 E_0}} \tag{5.21}$$

$$\beta_A = \begin{cases} 1 & \text{for} \quad \text{Class 1, 2, or 3 cross sections} \\ \dfrac{A_e}{A} & \text{for} \quad \text{Class 4 cross sections} \end{cases} \tag{5.22}$$

The values of the imperfection factor α and limiting slenderness $\overline{\lambda}_0$ can be obtained from Table 5.2 of the EC3 [5.19].

The effective area (A_e) is taken as the full area (A) for Class 1 $(D/t \leq 50\varepsilon^2)$, Class 2 $(D/t \leq 70\varepsilon^2)$, and Class 3 $(D/t \leq 90\varepsilon^2)$ cross sections, where ε is calculated as follows:

$$\varepsilon = \sqrt{\frac{235}{F_y} \frac{E_0}{210,000}} \tag{5.23}$$

It should be noted that the EC3 [5.19] does not provide design rules for the calculation of effective area (A_e) for Class 4 $(D/t > 90\varepsilon^2)$ slender circular hollow sections. In this study, the circular hollow sections investigated in the parametric study are classified as Class 4 slender sections, but no design provision is given in the EC3 for the calculation of the effective area. Hence, the full cross-sectional area (A) was used.

5.6.4 Proposed Design Equation

In this study, effective area equation for cold-formed stainless steel slender circular hollow section columns was proposed. The proposed effective area equation for Class 4 slender circular hollow sections is as follows:

$$A_e = A\varepsilon \left(\frac{125}{D/t}\right)^{0.1} \tag{5.24}$$

where A is the full cross-sectional area, ε is calculated from Eq. (5.23), D is the outer diameter, and t is the plate thickness of the circular stainless steel tube.

The proposed design strength for concentrically loaded cylindrical tubular compression members can be calculated in the same way as the EC3 [5.19]:

$$P_P = \chi F_y A_e \tag{5.25}$$

where χ is the reduction factor for flexural buckling that is calculated in the same way as in Eq. (5.19), F_y is the yield stress that is equal to the static 0.2% proof stress $(\sigma_{0.2})$, and A_e is the proposed effective area as given in Eq. (5.20).

5.6.5 Comparison of Column Strengths

The column strengths predicted from the parametric study were compared with the unfactored design strengths calculated using the American (ASCE [5.17]), Australian/New Zealand (AS/NZS [5.18]), and European (EC3 [5.19]) specifications for cold-formed stainless steel structures. The measured material properties obtained from the tensile coupon of Series C1, which is the same material properties as those used in the parametric study, were used to calculate the design strengths. The column strength ratios for all specimens are shown on the vertical axis of Figures 5.20−5.25, while the horizontal axis is plotted against the effective length (l_e) that is assumed equal to one-half of the column length. Figures 5.20 and 5.21 show that the design strengths calculated using the AS/NZS and EC3 specifications are unconservative for the columns having D/t ratios of 100 and 120, except for the short columns with lengths of 500 and 1000 mm. The design strengths calculated using the American Specification and the proposed design equation are generally conservative, except for some long columns. Figures 5.22−5.25 show that the design strengths calculated using the AS/NZS and EC3 specifications are

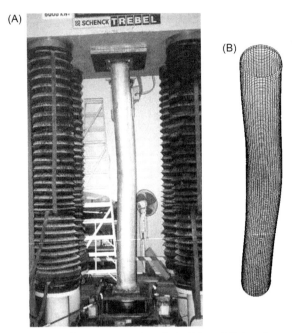

Figure 5.20 Comparison of experimental analysis (A) and finite element analysis (B) failure modes for specimen C2L2000 [5.16].

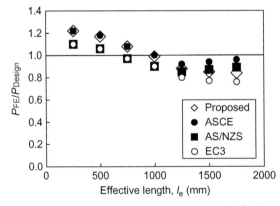

Figure 5.21 Comparison of FE strengths with design strengths for Series C100, [5.15].

generally unconservative for cold-formed stainless steel slender circular hollow section columns having D/t ratios of 140, 160, 180, and 200, while the ASCE is quite conservative. The design strengths predicted using the proposed design equation are generally conservative for

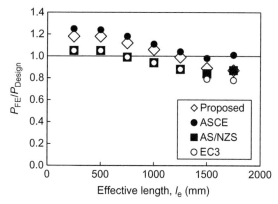

Figure 5.22 Comparison of FE strengths with design strengths for Series C120 [5.15].

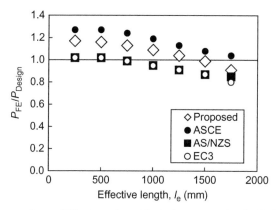

Figure 5.23 Comparison of FE strengths with design strengths for Series C140 [5.15].

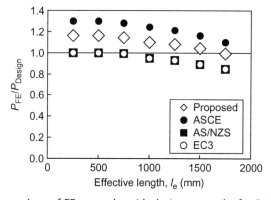

Figure 5.24 Comparison of FE strengths with design strengths for Series C160 [5.15].

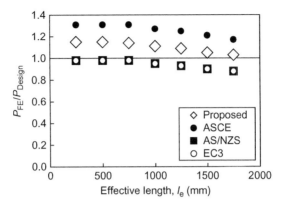

Figure 5.25 Comparison of FE strengths with design strengths for Series C180 [5.15].

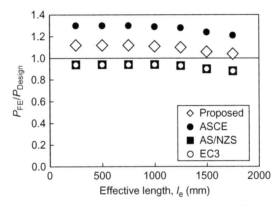

Figure 5.26 Comparison of FE strengths with design strengths for Series C200 [5.15].

cold-formed stainless steel slender circular hollow section columns having D/t ratios of 140, 160, 180, and 200.

REFERENCES

[5.1] Schmidt, H. Stability of steel shell structures. General report. Journal of Constructional Steel Research, 55(1), 159–181, 2000.

[5.2] Galambos, T. V. Guide to stability design criteria for metal structures. New York, NY: John Wiley & Sons, 1988.

[5.3] ANSYS Inc.. ANSYS 5.4-Manual: Theory, elements and commands references. USA, 1998.

[5.4] Schmidt, H. and Krysik, R. Towards recommendations for shell stability design by means of numerically determined buckling loads, In: Jullien, J. F., editor. Buckling of shell structures, on land, in the sea and in the air, Proc. Int. Coll. Lyon, London/New York: Elsevier Applied Science, 508–519, 1991.

[5.5]   EC3. Eurocode 3: Design of steel structures Part 1-6: Strength and stability of shell structures. London, UK: British Standards Institution, BS EN 1993-1-6, 2007.

[5.6]   Narayanan, S. and Mahendran, M. Ultimate capacity of innovative cold-formed steel columns. Journal of Constructional Steel Research, 59(4), 489–508, 2003.

[5.7]   Schafer, B. W. and Peköz, T. Computational modeling of cold-formed steel: Characterizing geometric imperfections and residual stresses. Journal of Constructional Steel Research, 47(3), 193–210, 1998.

[5.8]   Kwon, Y. B. and Hancock, G. J. A nonlinear elastic spline finite strip analysis for thin-walled sections. Thin-Walled Structures, 12(4), 295–319, 1991.

[5.9]   Raftoyiannis, I. G. and Ermopoulos, J. C. Stability of tapered and stepped steel columns with initial imperfections. Engineering Structures, 27(8), 1248–1257, 2005.

[5.10] EC3. Eurocode 3: Design of steel structures Part 1-1: General rules and rules for buildings. London, UK: British Standards Institution, BS EN 1993-1-1, 2005.

[5.11] Spyrakos, C. C. and Raftoyiannis, I. G. Linear and nonlinear finite element analysis. Pittsburgh, PA: Algor Publishing Division, 1998.

[5.12] Zhu, J. H. and Young, B. Aluminum alloy tubular columns—Part I: Finite element modeling and test verification. Thin-Walled Structures, 44(9), 961–968, 2006.

[5.13] Yan, J. and Young, B. Numerical investigation of channel columns with complex stiffeners—Part I: Tests verification. Thin-Walled Structures, 42(6), 883–893, 2004.

[5.14] Mazzolani, F. M. Aluminum alloy structures. 2nd ed., London: E & FN Spon, 1995.

[5.15] Young, B. and Ellobody, E. Column design of cold-formed stainless steel slender circular hollow sections. Steel and Composite Structures, 6(4), 285–302, 2006.

[5.16] Ellobody, E. and Young, B. Investigation of cold-formed stainless steel non-slender circular hollow section columns. Steel and Composite Structures, 7(4), 321–337, 2007.

[5.17] ASCE. Specification for the design of cold-formed stainless steel structural members. Reston, Virginia: American Society of Civil Engineers, SEI/ASCE-02, 2002.

[5.18] AS/NZS. Cold-formed stainless steel structures. Australian/New Zealand Standard. Sydney, Australia: Standards Australia, AS/NZS 4673:2001, 2001.

[5.19] EC3. Eurocode 3: Design of steel structures—Part 1.4: General rules—Supplementary rules for stainless steels. CEN, Brussels: European Committee for Standardization, ENV 1993-4, 1996.

[5.20] Young, B. and Hartono, W. Compression tests of stainless steel tubular members. Journal of Structural Engineering, ASCE, 128(6), 754–761, 2002.

[5.21] L. Gardner, A new approach to structural stainless steel design. PhD thesis. London: Department of Civil and Environmental Engineering, Imperial College of Science, Technology and Medicine, 2002.

[5.22] Ellobody, E. Buckling analysis of high strength stainless steel stiffened and unstiffened slender hollow section columns. Journal of Constructional Steel Research, 63(2), 145–155, 2006.

[5.23] Zhang, Y., Wang, C. and Zhang, Z. Tests and finite element analysis of pin-ended channel columns with inclined simple edge stiffeners. Journal of Constructional Steel Research, 63(3), 383–395, 2007.

[5.24] Becque, J. and Rasmussen, K. J. R. A numerical investigation of local-overall interaction buckling of stainless steel lipped channel columns. Journal of Constructional Steel Research, 65(8-9), 1685–1693, 2009.

[5.25] Gao, L., Sun, H., Jin, F. and Fan, H. Load-carrying capacity of high-strength steel box-sections I: Stub columns. Journal of Constructional Steel Research, 65(4), 918–924, 2009.

[5.26]   AISI. Specification for the design of cold-formed steel structural members. Washington, DC, 1996.

[5.27]   Theofanous, M. and Gardner, L. Testing and numerical modelling of lean duplex stainless steel hollow section columns. Engineering Structures, 31(12), 3047−3058, 2009.

[5.28]   Gardner, L. and Nethercot, D. A. Numerical modeling of stainless steel structural components: A consistent approach. Journal of Structural Engineering ASCE, 130 (10), 1586−1601, 2004.

[5.29]   Ashraf, M., Gardner, L. and Nethercot, D. A. Finite element modelling of structural stainless steel cross-sections. Thin-Walled Structures, 44(10), 1048−1062, 2007.

[5.30]   Goncalves, R. and Camotim, D. Geometrically non-linear generalised beam theory for elastoplastic thin-walled metal members. Thin-Walled Structures, 51, 121−129, 2012.

[5.31]   Goncalves, R. and Camotim, D. Generalised beam theory-based finite elements for elastoplastic thin-walled metal members. Thin-Walled Structures, 49(10), 1237−1245, 2011.

[5.32]   Zhu, J. H. and Young, B. Test and design of aluminum alloys compression members. Journal of Structural Engineering, 132(7), 1096−1107, 2006.

[5.33]   The Aluminum Association. Aluminum design manual. Washington, DC: The Aluminum Association, 2005.

[5.34]   AS/NZS. Aluminum structures—Part 1: Limit state design, Australian/New Zealand Standard AS/NZS 1664.1:1997. Sydney, Australia: Standards Australia, 1997.

[5.35]   EC9. Eurocode 9: Design of aluminum structures—Part 1-1: General rules— General rules and rules for buildings. Final Draft October 2000. European Committee for Standardization, DD ENV 1999-1-1:2000, 2000.

[5.36]   Young, B. Tests and design of fixed-ended cold-formed steel plain angle columns. Journal of Structural Engineering, ASCE, 130(12), 1931−1940, 2004.

[5.37]   Australian Standard. Steel sheet and strip—hot-dipped zinc-coated or aluminum/ zinc-coated, AS 1397. Sydney, Australia: Standards Association of Australia, 1993.

[5.38]   Australian Standard. Methods for tensile testing of metals, AS 1391. Sydney, Australia: Standards Association of Australia, 1991.

[5.39]   American Iron and Steel Institute (AISI). Specification for the design of cold-formed steel structural members. Washington, DC: AISI, 1996.

[5.40]   Australian/New Zealand Standard (AS/NZS). Cold-formed steel structures. Sydney, Australia: Standards Australia, AS/NZS 4600:1996, 1996.

[5.41]   Young, B. and Rasmussen, K. J. R. Tests of fixed-ended plain channel columns. Journal of Structural Engineering, ASCE, 124(2), 131−139, 1998.

[5.42]   Rasmussen, K. J. R. and Rondal, J. Strength curves for metal columns. Journal of Structural Engineering ASCE, 113(6), 721−728, 1997.

CHAPTER 6

Examples of Finite Element Models of Metal Beams

6.1. GENERAL REMARKS

This chapter presents examples of different finite element models of metal beams based on the background of finite element analysis as detailed in Chapters 1−4. The examples presented in this chapter are already published in journal papers. The examples are arbitrary, chosen from research conducted by the authors of this book so that all related information regarding the finite element models developed in the papers can be provided to readers. The chosen finite element models are for beams constructed from different metals having different mechanical properties, cross sections, boundary conditions, and geometries. Once again, when presenting the previously published models developed for metal beams, the main objective is not to repeat the previously published information but to explain the fundamentals of the finite element method used in developing the models.

This chapter starts with a review of recently published finite element models on metal beams. After that, the chapter presents three examples of finite element models on metal beams previously published by the authors. The authors also highlight how the information presented in the previous chapters is used to develop the examples of finite element models as discussed in the current chapter. The experimental investigations simulated, finite element models developed, verifications of finite element models, results obtained, and comparisons with design values in current specifications are presented in this chapter with clear references. The authors have an aim that the presented examples highlighted in this chapter can explain to readers the effectiveness of finite element models in providing detailed data that augment experimental investigations conducted on metal beams. The results are discussed to show the significance of the finite element models in predicting the structural response of metal beams.

Finite Element Analysis and Design of Metal Structures
DOI: http://dx.doi.org/10.1016/B978-0-12-416561-8.00006-8
115

6.2. PREVIOUS WORK

Many finite element models were developed in the literature. Some of these models with detailed examples are presented in the following paragraphs for the investigations of the behavior and design of metal beams. The aforementioned numerical investigations were performed on metal beams constructed from steel, cold-formed steel, stainless steel, and aluminum alloy materials. Liu and Chung [6.1] have investigated the structural performance of steel beams with large web openings and various shapes and sizes through finite element analysis. The study has looked into the Vierendeel mechanism in steel beams with web opening and the empirical interaction formulae presented in current codes of practice. The finite element analysis detailed in the study [6.1] included steel beams with web openings of various shapes and sizes. The study has shown that all steel beams with large web openings of various shapes behave similarly under a wide range of applied moments and shear forces. It was also shown that the failure modes were common in all beams and the yield patterns of those perforated sections at failure are similar. The authors have performed parametric studies using finite element method, which yielded a simple empirical design method applicable for perforated sections with web openings of various shapes and sizes. The authors have used isoparametric 8-node shell element (see Section 3.2). The stresses incorporated in each shell element include two in-plane direct stresses, one in-plane shear stress, and two out-of-plane shear stresses. The material nonlinearity (see Section 4.4) was incorporated into the finite element model. A bilinear stress—strain curve was adopted in the material modeling of steel together with the von Mises yield criteria and nonassociate plastic flow rule (see Section 3.4). It should be noted that in order to model load redistribution after yielding, elastic unloading was also incorporated. The geometric nonlinearity (see Section 4.5) was also incorporated into the finite element model, with large deformation in the perforated section after yielding may be predicted accurately to allow for load redistribution within the perforated sections. This allowed the Vierendeel mechanism with the formation of plastic hinges in the tee-sections above and below the web openings to be investigated in detail. The authors have found that the finite element modeling produced good comparisons with laboratory tests on steel beams with circular web openings having diameter of 0.60 times the overall height of the beam (h), which is subjected to significant moment—shear interaction [6.2,6.3]. It was concluded that the finite element models are considered to be applicable in the present study to perforated sections with various shapes and sizes. The developed finite element model [6.1] was used to perform parametric studies.

Mohri et al. [6.4] have performed a combined theoretical and numerical stability analyses of unrestrained mono-symmetric thin-walled beams. The study discussed a contribution to the overall stability of unrestrained thin-walled elements with open sections. The authors have developed a nonlinear finite element model for beam lateral buckling stability analysis. The authors have looked into the EC3 [5.10] approach and performed analytical solutions for checking the stability of laterally unrestrained beams. The analytical solutions were discussed in the study [6.4] and compared against finite element results. The general-purpose finite element software ABAQUS [1.27] was used in numerical simulations. Three-dimensional beam elements with warping (B31OS) and shell elements (S8R5) (see Section 3.2) were chosen in modeling the lateral buckling phenomena. The authors have found that the analytical results were close to shell element results. The authors have found that shell elements could consider local buckling, section distortion, and the local effects of concentrated loads and boundary conditions, which were all ignored in the beam element theory.

Zhu and Young [6.5] have conducted a combined experimental and numerical investigation of aluminum alloy flexural members. The investigations were performed on different sizes of square hollow sections subjected to pure bending. Material properties of each specimen were also measured. The authors have developed a nonlinear finite element model, which was verified against the pure bending tests. The verified finite element model was used to perform parametric studies on aluminum alloy beams of square hollow sections. The experimental and numerical bending strengths were compared against the design strengths calculated using the American [5.33], Australian/New Zealand [5.34], and European [5.35] specifications for aluminum structures. The bending strengths were also compared with the design strengths predicted by the direct strength method, which was developed for cold-formed carbon steel members. The authors have proposed design rules for aluminum alloy square hollow section beams based on the current direct strength method. In addition, reliability analysis was performed to evaluate the reliability of the design rules. The general-purpose finite element software ABAQUS [1.27] was used in the analysis for the simulation of aluminum alloy beams subjected to pure bending. Residual stresses were not included in the model. This is because in extruded aluminum alloy profiles, residual stresses have very small values. For practical purpose, these have a negligible effect on load-bearing capacity as recommended by Mazzolani [5.14]. The authors have modeled only one half of the beam due to symmetry (see Section 3.3). The displacement control loading method was used, which is identical to that used in the beam tests. The load was applied in increments using the Riks method (see Section 4.6) with automatic increment size being applied. The material nonlinearity or "plasticity" was included in the finite element model. The general-purpose shell elements (S4R) were used in the finite element model. The authors have conducted convergence studies to choose the finite element mesh that provides accurate results with minimum computational time. It was found that when meshing each side of the cross section with 10 elements, the results will be accurate compared with the tests.

Liu and Gannon [6.6] have presented the results of a finite element study on the behavior and capacity of W-shape steel beams reinforced with welded plates under loading. The authors have developed a nonlinear finite element model, which was verified against published test results by one of the authors [6.7]. The verified model was used to conduct parametric studies investigating the effects of reinforcing patterns, preload magnitudes at the time of welding, and initial imperfections of the unreinforced beam. The study [6.6] has shown that the increase in the preload magnitude at the time of reinforcing resulted in a reduction in the ultimate capacity of reinforced beams failing by lateral torsional buckling (LTB). The finite element model was developed using the general-purpose finite element software ANSYS [5.3] to simulate a flexural member under a four-point loading with reinforcing plates added at various load levels. The beam cross section, weld, and reinforcing plates were all modeled using Shell 181 element as specified in ANSYS [5.3]. This 3D, 4-node element is suitable for the analysis of thin to moderately thick structures with large rotation and large strain nonlinearities and also capable of buckling simulation (see Section 3.2). Considering the symmetry of specimen geometry and loading, only half of a specimen was modeled. Convergence studies were carried out by the authors to choose the best finite element mesh. It was found that a mesh size of 10×20 mm was selected for flanges and reinforcing plate and a mesh size of 15×20 mm was used for the web. Bearing plates at the loading points and end supports were modeled using the Solid 45 brick element to allow the load to be distributed across the width of the flanges. Solid 45 is a 3D, 8-node structural solid element capable of including effects of large deflection, large strain, and stress stiffening (see Section 3.2). The effect of initial geometric imperfections on unreinforced steel beams was investigated in the study [6.6]. Also, residual stresses were incorporated into the finite element models at two different stages. The initial residual stress for the beam section alone was transformed into a series of discrete uniform stresses that could be applied to each element in the model. These residual stresses were then specified by creating an initial stress field applied to the model during the first substep of the first load step of the analysis. At the second stage of the analysis, the residual stresses in the reinforcing plates and in the beam section due to welding were added after the beam had reached the specified preload magnitude. Since the initial stress file may only be input in the first load step, residual stresses at the second stage were introduced alternatively by specifying a temperature body force on the shell elements. A similar technique was used by Wu and Grondin [6.8]. In the nonlinear analysis, a Newton–Raphson procedure (see Chapter 4) was used to perform equilibrium iterations until convergence criteria were satisfied. Loads were applied to the model over a load step which was divided into a number of substeps to obtain an accurate solution.

Theofanous and Gardner [6.9] have reported material and three-point bending tests on lean duplex stainless steel hollow section beams. The three-point bending tests were simulated by finite element analysis. The validated model was used to perform parametric studies to assess the effects of cross section aspect ratio, cross section slenderness, and moment gradient on the strength and deformation capacity of lean duplex stainless steel beams. Based on both the experimental and numerical results, the authors have proposed slenderness limits and design rules for incorporation into structural stainless steel design standards. The authors have used the general-purpose finite element software ABAQUS [1.27] to develop the models. The initial geometric imperfections, material properties, and mesh density were investigated by the authors. The reduced integration 4-node doubly curved general-purpose shell element S4R with finite membrane strains (see Section 3.2) has been used to simulate the structural behavior of the lean duplex stainless steel beams. The element has been previously used by the authors [6.10,6.11] and shown to perform well in the modeling of thin-walled metallic structures. Mesh convergence studies (see Section 3.3) were performed by the authors to evaluate the best mesh that provides accurate results with reasonable computational time. Only half of the cross section of each specimen was modeled due to symmetry. The load was applied as a point load at the junction of the web with the corner radius in the lower (tension) part of the beam to avoid web crippling. The authors have used the measured geometry and material properties in the finite element models. Due to the absence of global buckling, only local geometric imperfections have been incorporated in the finite element models in the form of the lowest buckling mode shape. A linear eigenvalue buckling analysis (see Section 4.3) was therefore initially conducted using the subspace iteration method for eigenmode extraction. Subsequently, a geometrically and materially nonlinear analysis (see Sections 4.4 and 4.5) incorporating geometric imperfections was carried out. The modified Riks method (see Section 4.6) was employed in the nonlinear analyses.

Sweedan [6.12] has numerically investigated the lateral stability of cellular steel beams. Three-dimensional finite element modeling was presented for simply supported I-shaped cellular steel beams having different cross-sectional dimensions, span lengths, and web perforation configurations. The author has performed stability analyses for beams subjected to equal end moments, mid-span concentrated loads, and uniformly distributed loads. The study has shown that the moment gradient coefficient was considerably affected by the beam geometry, slenderness, and web perforation configuration. Based on the study [6.12], a simplified approach was developed to enable accurate prediction of a moment modification factor for cellular beams. It was shown that the proposed approach allows for accurate and conservative evaluation of the critical moment associated with the lateral torsional−distortional buckling of cellular beams. The study also has presented several numerical examples to illustrate the application of the proposed procedure. The 3D finite element model for I-shaped cellular steel beams was developed using the general-purpose finite element software ANSYS [5.3]. The 4-node shell element (Shell 181), which has six degrees of freedom at each node including three translations and three rotations, was used. Numerical computations were conducted for cellular steel beams that were assumed to be constructed of linear elastic material. Convergence studies were performed to choose the best finite element mesh that provides accurate results with less computational time.

Haidarali and Nethercot [6.13] have developed two series of finite element models investigating the buckling behavior of laterally restrained cold-formed steel Z-section beams. The developed models have accounted for the nonlinear material and geometry. The first series of the models allowed for the possibility of combined local–distortional buckling, and the second series of models allowed for local buckling only. The authors have used previously published four-point bending tests to verify the finite element models. The general-purpose finite element software ABAQUS [1.27] has been used to perform the nonlinear analyses. The finite element results compared well against the tests. The general-purpose shell elements S4R were used for all the components of the finite element models. The initial geometric imperfections were included in the models. The authors have performed an elastic eigenvalue buckling analysis to obtain appropriate eigenmodes for local and distortional buckling. These buckling modes were inserted into the nonlinear analysis to define the shape and distribution of initial imperfections. The authors have found some difficulties for the local buckling mode for some sections as the pure local buckling modes for these sections were obtained at very high eigenmodes, which required considerable computational time. Therefore, it was decided to generate the shape and distribution of initial imperfections manually with the aid of the finite strip software CUFSM [6.14]. The classical finite strip method uses polynomial functions for the deformed shape in the transverse direction, while a single half sine wave is used for the longitudinal shape function.

Ellobody [6.15] has investigated the behavior of normal and high strength castellated steel beams under combined lateral torsional and distortional buckling modes. An efficient nonlinear 3D finite element model has been developed for the analysis of the beams. The initial geometric imperfections and material nonlinearities were carefully considered in the analysis. The nonlinear finite element model was verified against tests on castellated beams having different lengths and different cross sections. Failure loads and interaction of buckling modes as well as load–lateral deflection curves of castellated steel beams were investigated in this study. The author has performed parametric studies to investigate the effects of the cross section geometries, beam length, and steel strength on the strength and buckling behavior of castellated steel beams. The study [6.15] has shown that the presence of web distortional buckling (WDB) causes a considerable decrease in the failure loads of slender castellated steel beams. It was also shown that the use of high strength steel offers a considerable increase in the failure loads of less slender castellated steel beams. The failure loads predicted from the finite element model were compared with that predicted from Australian Standards [6.16] for steel beams under LTB. It was shown that the Specification predictions are generally conservative for normal strength castellated steel beams failed by LTB, but unconservative for castellated steel beams failed by WDB, and quite conservative for high strength castellated steel beams failed by LTB. The general-purpose finite element software ABAQUS [1.27] was used to perform the finite element analyses. The author has performed an eigenvalue buckling analysis, which was followed by a nonlinear load−displacement analysis. The material and geometric nonlinearities were included in the analyses. The author has used a combination of 4-node and 3-node doubly curved shell elements with reduced integration (S4R and S3R). Convergence studies were performed to choose the finite element mesh that provides accurate results with minimum computational time. It was found that approximately 15×15 mm (length by width of S4R element and depth by width of S3R element) provides adequate accuracy in modeling the web, while a finer mesh of approximately 8×15 mm was used in the flange. The initial geometric imperfections were included in the analyses.

Ellobody [6.17] has extended the study [6.15] to discuss the nonlinear analysis of normal and high strength cellular steel beams under combined buckling modes. The author has developed a nonlinear 3D finite element model, which accounted for the initial geometric imperfections, residual stresses, and material nonlinearities of flange and web portions of cellular steel beams. The nonlinear finite element model was verified against tests on cellular steel beams having different lengths, cross sections, loading conditions, and failure modes. Failure loads, load−mid-span deflection relationships, and failure modes of cellular steel beams were predicted from the finite element analysis. The author has performed parametric studies involving 120 cellular steel beams using the verified finite element model to study the effects of the cross section geometries, beam length, and steel strength on the strength and buckling behavior of cellular steel beams. The study [6.17] has shown that cellular steel beams failed by combined web distortional and web-post buckling (WPB) modes exhibited a considerable decrease in the failure loads. It was also shown that the use of high strength steel offers a considerable increase in the failure loads of less slender cellular steel beams. The failure loads predicted from the finite element model were compared with that predicted from Australian Standards [6.16] for steel beams under LTB. It was shown that the specification predictions are generally conservative for normal strength cellular steel beams failed by LTB, but unconservative for cellular steel beams failed by combined web distortional and WPB, and quite conservative for high strength cellular steel beams failed by LTB. The general-purpose finite element software ABAQUS [1.27] was used to perform the finite element analyses. The author has performed an eigenvalue buckling analysis, which was followed by a nonlinear load—displacement analysis. Both material and geometric nonlinearities were included in the analyses. The author has used a combination of 4-node and 3-node doubly curved shell elements with reduced integration (S4R and S3R). Convergence studies were performed to choose the finite element mesh that provides accurate results with minimum computational time. It was found that approximately 40×50 mm (length by width of S4R element and depth by width of S3R element) provides adequate accuracy in modeling the web, while a mesh of approximately 35×50 mm was used in the flange. The initial geometric imperfections and residual stresses were included in the analyses.

Anapayan and Mahendran [6.18] have presented the performance of LiteSteel Beam (LSB), which is a new hollow flange channel section developed using a patented dual electric resistance welding and cold-forming process. The beam has a unique geometry consisting of torsionally rigid rectangular hollow flanges and a slender web, and is commonly used as flexural members. The authors have shown that the LSB flexural members are subjected to a relatively new lateral distortional buckling mode, which reduces their moment capacities. Therefore, a detailed investigation into the lateral buckling behavior of LSB flexural members was undertaken by the authors using experiments and finite element analyses. This study has presented the finite element models developed to simulate the behavior and capacity of LSB flexural members subjected to lateral buckling. The models have included material inelasticity, lateral distortional buckling deformations, web distortion (WD), residual stresses, and geometric imperfections. The study [6.18] included a comparison of the finite element ultimate moment capacities with predictions from other numerical analyses and available buckling moment equations as well as experimental results. It was shown that the developed finite element models accurately predicted the behavior and moment capacities of LSBs. The validated model was then used to perform parametric studies, which produced accurate moment capacity data for all the LSB sections and improved design rules for LSB flexural members subjected to lateral distortional buckling. The general-purpose finite element software ABAQUS [1.27] was used to perform the analyses. The shell element S4R5 was used to develop the LSB model. Convergence studies were performed to predict the best finite element mesh that provides accurate results with less computational time. It was shown that a minimum mesh size density comprising 5×10 mm elements was required to represent accurate residual stress distributions, spread of plasticity, and local buckling deformations of LSBs. Element widths less than or equal to 5 mm and a length of 10 mm were selected as the suitable mesh size. Nine integration points through the thickness of the elements were used to model the distribution of flexural residual stresses in the LSB sections and the spread of plasticity through the thickness of the shell elements, as recommended by Kurniawan and Mahendran [6.19].

Zhou et al. [6.20] have investigated the performance of aluminum alloy plate girders subjected to shear force through numerical investigation. The aluminum alloy plate girders were fabricated by welding of three plates to form an I-section. The authors have developed a nonlinear finite element model, which was verified against experimental results. The geometric and material nonlinearities were included in the finite element model. The welding of the aluminum plate girders and the influence of the heat-affected zone (HAZ) were included in the finite element model. The ultimate loads, web deformations, and failure modes of aluminum plate girders were predicted from the study. The authors have performed parametric studies investigating the effects of cross section geometries and the web slenderness on the behavior and shear strength of aluminum alloy plate girders. The shear resistances obtained from the study [6.20] were compared against the design strengths predicted using the European Code [5.35] and American Specifications [5.33] for aluminum structures. Based on the study, the authors have proposed a design method to predict the shear resistance of aluminum alloy plate girders by modifying the design rules specified in the European Code [5.35]. The general-purpose finite element software ABAQUS [1.27] was used in the analysis for the simulation of aluminum alloy plate girders subjected to shear force. The material nonlinearity or plasticity was included in the finite element model. The general-purpose shell elements S4R were used in the finite element model. The authors have conducted convergence studies to choose the finite element mesh that provides accurate results with minimum computational time. It was found that a mesh size of 8×8 mm (length by width) in the HAZ area and a mesh size of 15×15 mm elsewhere provided accurate results compared with the tests.

Soltani et al. [6.21] have developed a numerical model to predict the behavior of castellated beams with hexagonal and octagonal openings.

The material and geometric nonlinearities were considered in the model. The authors have performed an eigenvalue buckling analysis to model the initial geometric imperfections. The finite element model was verified against previously published experimental results. The ultimate loads and general failure modes were predicted from the study. The numerical results have been compared with those obtained from the design method presented in the EC3 [5.10]. The general-purpose finite element software LUSAS [6.22] was used to perform the finite element analyses. The webs, flanges, intermediate plates, and stiffeners of the castellated steel beams were modeled by a 3D 8-node thin shell element QSL8 available in the LUSAS [6.22] element library. The element has three translational degrees of freedom at each of four corners and four mid-sides and normal rotations at the two Gauss points along each side. Only one layer of elements was used through thickness direction. Finer meshes were generated to model areas near the openings in order to improve precision and accommodate the opening shapes. Only half of the beam was modeled owing to symmetry to reduce the model size and subsequent processing time. Although the cross section was also symmetrical about its major and minor axes, it was necessary to model the full cross section because the buckled shape is nonsymmetrical. The regular meshing was employed for all components of the beam. The density and the configuration of the finite element mesh were determined based on results obtained from convergence studies in order to provide a reasonable balance between accuracy and computational expense. The dimensions of elements with the minimum width, located around the opening, were chosen to ensure the aspect ratio was kept below 5. The nonlinear analyses were performed with meshes varying from 1472 to 3776 elements. The geometric nonlinearity was considered to account for the large displacements. The geometrically nonlinear analysis followed the continually changing geometry of the beam when formulating each successive load increment.

Kankanamge and Mahendran [6.23] have highlighted the importance of cold-formed steel beams as floor joists and bearers in buildings. The authors have undertaken finite element analyses to investigate the LTB behavior of simply supported cold-formed steel lipped channel beams subjected to uniform bending. The general-purpose finite element software ABAQUS [1.27] was used to develop the finite element model, which was verified using available numerical and experimental results. The validated finite element model was used to perform parametric studies to simulate the LTB behavior and capacity of cold-formed steel beams under varying conditions. The authors have compared the moment capacity results against the predictions from the current design rules in many cold-formed steel codes, and suitable recommendations were made. The authors have found that European design rules [6.24] were conservative, while Australian/New Zealand [6.25] and North American [6.26] design rules were unconservative. Therefore, the authors have proposed design equations for the calculation of the moment capacity based on the available finite element analysis results. This study [6.23] has presented the details of the parametric study, recommendations to the current design rules, and the new design rules proposed in this research for LTB of cold-formed steel lipped channel beams.

6.3. FINITE ELEMENT MODELING AND RESULTS OF EXAMPLE 1

The first example presented in this chapter is the simply supported castellated beams constructed from carbon steel modeled by Ellobody [6.15]. The castellated steel beams were subjected to distortional buckling and tested by Zirakian and Showkati [6.27] (Figure 6.1). The castellated

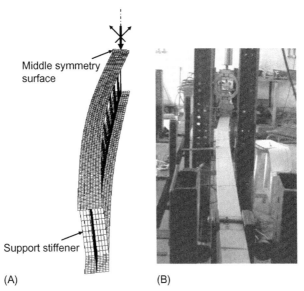

(A)   (B)

Figure 6.1 Comparison of experimental (B) and numerical (A) buckled shapes at failure for castellated steel beam specimen having a depth of 210 mm and a length of 4400 mm [6.15,6.27].

beams were loaded with central concentrated load. Lateral deflections were prevented at mid-span and near the supports using lateral bracing. The testing program included six full-scale beam tests having different cross section geometries and lengths as detailed in Ref. [6.27]. The castellated beam tests were designed so that the top compression flange of the beam was restrained against lateral buckling at mid-span and near the supports. Hence, the cross section at quarter-span was subjected to unrestrained distortional buckling, while the cross section at mid-span was subjected to restrained distortional buckling. The two buckling modes are detailed in [6.27,6.28]. The load was applied step-by-step until failure occurred. Failure was identified when the lateral deflections were large at quarter-span locations and unloading took place. The material properties of flange and web portions were determined from tensile coupon tests as detailed in Ref. [6.27]. The test program provided useful and detailed data regarding the behavior of castellated steel beams and conformed to the criteria of a successful experimental investigation highlighted in Section 1.3.

The tests carried out by Zirakian and Showkati [6.27] were modeled by Ellobody [6.15] using the general-purpose finite element software ABAQUS [1.27]. The model has accounted for the measured geometry, initial geometric imperfections, and measured material properties of flange and web portions. The finite element analyses of the castellated steel beams comprised linear eigenvalue buckling analysis (see Section 4.3) as well as materially and geometrically nonlinear analyses detailed in Sections 4.4 and 4.5, respectively. A combination of 4-node and 3-node doubly curved shell elements with reduced integration (S4R and S3R) were used to model the flanges and web of the castellated steel beams (see Section 3.2). In order to choose the finite element mesh that provides accurate results with minimum computational time, convergence studies (see Section 3.3) were conducted. It was found that approximately 15×15 mm (length by width of S4R element and depth by width of S3R element) provides adequate accuracy in modeling the web, while a finer mesh of approximately 8×15 mm was used in the flange. Only half of the castellated beam was modeled due to symmetry (see Section 3.3). The load was applied in increments as concentrated static load, which was also identical to the experimental investigation.

The stress—strain curve for the structural steel given in the EC3 [5.10] was adopted in the study [6.15] with measured values of the yield stress and ultimate stress used in the tests [6.27]. The material behavior provided by ABAQUS [1.27] (using the PLASTIC option) allows a nonlinear stress—strain curve to be used (see Section 3.4). Buckling of castellated beams depends on the lateral restraint conditions to compression flange and geometry of the beams. Two main buckling modes detailed in Refs [6.27,6.28] could be identified as unrestrained and restrained lateral distortional buckling modes. The lateral distortional buckling modes could be obtained by performing eigenvalue buckling analysis (see Section 4.3) for castellated beams with actual geometry and lateral restraint conditions to the compression flange. Only the first buckling mode (eigenmode 1) was used in the eigenvalue analysis. Since buckling modes predicted by ABAQUS eigenvalue analysis [1.27] are generalized to 1.0, the buckling modes were factored by a magnitude of $L_u/1000$, where L_u is the length between points of effective bracing. The factored buckling modes were inserted into the load—displacement nonlinear analysis of the castellated beams following the eigenvalue prediction. It should be noted that the investigation of castellated beams with different slenderness ratios could result in LTB mode with or without WDB mode. Hence, the eigenvalue buckling analysis must be performed for each castellated beam with actual geometry to ensure that the correct buckling mode could be incorporated in the nonlinear displacement analysis.

The finite element model for castellated beams under distortional buckling developed by Ellobody [6.15] was verified against the test results detailed in Ref. [6.27]. The failure loads, failure modes, and load—lateral deflection curves obtained experimentally and numerically using the finite element model were compared. The failure loads obtained from the tests as well as calculated using the design equation proposed in Ref. [6.29] as reported in Ref. [6.27] ($P_{Test/Calculated}$) and finite element analyses performed in Ref. [6.15] (P_{FE}) were compared. The mean value of $P_{Test/Calculated}/P_{FE}$ ratio is 1.01 with the coefficient of variation (COV) of 0.020. Three failure modes were observed experimentally [6.27] and verified numerically [6.15] using the finite element model. All the tested castellated beams [6.27] underwent LTB and WD, while steel beam yielding (SY) was observed in castellated beams with lengths of 3600 and 4400 mm. The SY failure mode was predicted from the finite element model by comparing the von Mises stresses in the castellated beams at failure against the measured yield stresses. On the other hand, the SY failure mode was judged in the tests by comparing the test failure loads against the plastic collapse loads (P_{px}) calculated according to AS4100 [6.16]. The load—lateral deflection curves predicted experimentally and numerically were compared as shown in Figure 6.2. The curves

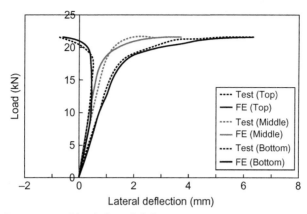

Figure 6.2 Comparison of load–lateral deflection curves at quarter-span of castellated steel beam specimen having a depth of 180 mm and a length of 3600 mm [6.15].

were plotted as an example at quarter-span of test specimen C180-3600 at the top, middle, and bottom points of the web of castellated beam. It was shown that good agreement was generally achieved between experimental and numerical results. The positive sign represents the lateral deflection in front of the web and the negative sign represents the lateral deflection at back of the web. The deformed shapes of castellated beams at failure observed experimentally and numerically were also compared. Figure 6.1 showed as an example of the buckled shape of specimen C210-4400 observed in the test in comparison with that predicted from the finite element analysis. It was shown that the experimentally and numerically deformed shapes are in good agreement. The failure mode observed experimentally and confirmed numerically was a combination of LTB, WD, and SY. The data obtained from ABAQUS [1.27] has shown that the von Mises stresses at the maximum stressed fibers at the top and bottom flanges at mid-span exceeded the measured yield stresses.

The verified finite element model developed in Ref. [6.15] was used to study the effects of the cross section geometries, beam length, steel strength, and nondimensional slenderness on the strength and buckling behavior of castellated steel beams. A total of 96 castellated steel beams were analyzed using the finite element model. The dimensions and material properties of the castellated steel beams were reported in Ref. [6.15]. The investigated castellated steel beams had different nondimensional slenderness (λ) calculated based on AS4100 [6.16] ranged from 1.0 to 3.1. The nondimensional slenderness (λ) is equal to the square root of the major axis full plastic moment divided by the elastic buckling moment, which was considered as a guide for beam slenderness in the study. The failure loads and failure modes of the castellated steel beams were predicted from the parametric studies as reported in Ref. [6.15]. The study [6.15] has shown that the failure loads of the castellated beams showed logical and expected results, with less slender beams followed a more "plastic collapse" mode, which are obviously driven by the steel strength. The more slender the beam, the more elastic buckling will be obtained. Furthermore, the collapse behavior is dependent on the lateral torsional and WDB behavior of the beam. It was also shown that the use of high strength steel offered a considerable increase in the failure loads of less slender castellated steel beams.

The plastic collapse load (P_{px}) and the design failure load (P_{AS4100}) calculated according to AS4100 [6.16] were obtained in the study [6.15]. The failure loads obtained from the parametric study (P_{FE}) were compared against the design failure loads calculated using the AS4100 [6.16] (P_{AS4100}) for the castellated steel beams. The study [6.15] has shown that the specification predictions were generally conservative for the castellated beams failed by LTB and having steel yield stress of 275 MPa. The specification predictions were unconservative for the beams failed mainly by WD. The specification predictions were also unconservative for the beams failed by combined (LTB + SY + WD) and (LTB + WD). On the other hand, the specification predictions were quite conservative for all remaining castellated steel beams, particularly those fabricated from high strength steel. The failure loads of castellated steel beams predicted from the finite element analysis (P_{FE}) and design guides (P_{AS4100}) were nondimensionlized with respect to the plastic collapse load (P_{px}) and plotted against the nondimensional flange width-to-thickness ratio (B/t) in Figure 6.3 as reported in Ref. [6.15]. The comparison of the numerical and design predictions has shown that the AS4100 design guide was generally conservative for the castellated steel beams with normal yield stresses, while it was quite conservative for the beams with higher yield stresses.

6.4. FINITE ELEMENT MODELING AND RESULTS OF EXAMPLE 2

The second example presented in this chapter is the beams constructed from aluminum alloy carried out by Zhu and Young [6.5], which provided the experimental bending strengths, moment–curvature curves, and failure modes of the aluminum alloy beams. A nonlinear finite element model was developed to simulate the flexural behavior of the beams based on the test results as described in Ref. [6.5]. The tested beams [6.5] had square hollow sections and were subjected to pure bending (Figure 6.4). The test specimens were fabricated by extrusion using high strength 6061-T6 heat-treated aluminum alloys. The test program included 10 simply supported beams, which is detailed in Ref. [6.5] and, once again, no intention to repeat the materials published in this paper. However, it should be mentioned that the experimental program presented was planned such that 10 beam tests were accurately investigated. The tests were well instrumented such that the experimental results were used in the verification of the finite element models developed by Zhu and Young [6.5]. Material properties of each specimen were determined by longitudinal tensile coupon tests. Hinge and roller supports were simulated by half round and pin, respectively. The simply supported specimens were loaded symmetrically at two points to the bearing plates within the moment span using a spreader beam. Half round and pin were also used at the loading points. Stiffening plates were used at the loading points and supports to prevent web bearing failure at the load of concentration. The experimental ultimate moments (M_{Exp}) were obtained using half of the ultimate applied load from the actuator multiplied by the lever arm (distance from the support to the loading point) of the specimens.

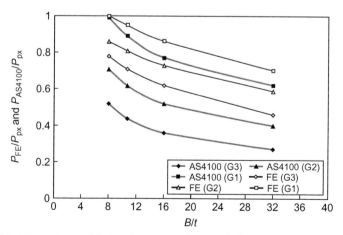

Figure 6.3 Comparison of finite element analysis and design predictions for castellated steel beams in groups G1–G3 having a depth of 180 mm and a length of 5200 mm with different steel grades [6.15].

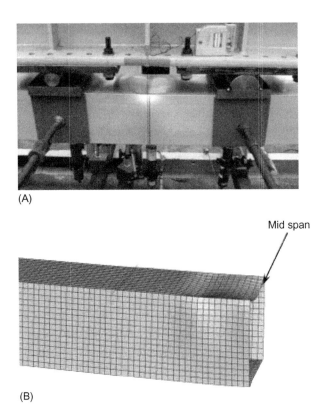

(A)

(B)

Figure 6.4 Comparison of experimental (A) and finite element (B) analysis failure modes of aluminum alloy beam of square hollow section $153 \times 153 \times 3$ [6.5].

The mass of the spreader beam, half round, pin, bearing plates, and stiffening plates were included in the calculation of the ultimate moments. The observed failure modes included local buckling (L) and material yielding (Y).

The general-purpose finite element program ABAQUS [1.27] was used in the study [6.5] for the simulation of aluminum alloy beams subjected to pure bending. Residual stresses were not included in the model. Only half of the beam was modeled by using symmetric condition at mid-length of the specimen. The midpoint of the bottom flange at mid-length was restrained for longitudinal degree of freedom to avoid the singularity of the stiffness matrix. The displacement control loading method was used in the finite element model, which is identical to that used in the beam tests. The load was applied to the beam by specifying a displacement to the reference point of the rigid surface at the loading point. The default Riks method (see Section 4.6) with automatic increment size was applied. The measured material properties of the test specimens were used in the finite element model. The material nonlinearity or plasticity was included in the finite element model using a mathematical model known as the incremental plasticity model available in the ABAQUS [1.27] library. The model was based on the centerline dimensions of the cross sections. The general-purpose shell elements S4R (see Section 3.2) were used in the finite element model. In order to choose the finite element mesh that provides accurate results with minimum computational time, mesh sensitivity and convergence studies were carried out in this study. Generally, each cross section was meshed into 10 elements.

The finite element model developed by the authors [6.5] was verified against the experimental results presented in the same paper [6.5]. The finite element model was also verified against three pure bending test results reported by Zhu and Young [6.30]. The ultimate moments and failure modes predicted by the finite element analysis were compared with the experimental results [6.5,6.30]. It was shown that the ultimate moments (M_{FEA}) obtained from the finite element model are generally in good agreement with the experimental ultimate moments (M_{Exp}). The mean value of the experimental-to-numerical ultimate moment ratio (M_{Exp}/M_{FEA}) was 0.98 with the corresponding COV of 0.044, as presented in Ref. [6.5]. The failure modes at ultimate moment obtained from the tests and finite element analysis for each specimen were also compared in Ref. [6.5]. The observed failure modes included local buckling (L) and material yielding (Y) due to large deflection. The failure modes predicted by the finite element analysis were in good agreement with those observed in the tests. Figure 6.4A showed a photograph of specimen H-153 × 153 × 3 after the ultimate moment has been reached. The specimen failed by local buckling at mid-span. Figure 6.4B showed the deformed shape of the same specimen predicted by the finite element analysis after the ultimate moment has been reached. Figure 6.5 shows a comparison of the moment-curvature curves obtained from the test and predicted by the finite element analysis for specimen H-76 × 76 × 3. It was shown that the finite element analysis curve followed the experimental curve closely, except that the moments predicted by the finite element analysis were slightly higher than the experimental moments in the nonlinear region.

The verified finite element model was used to perform parametric studies, which included 60 beam specimens of square hollow sections

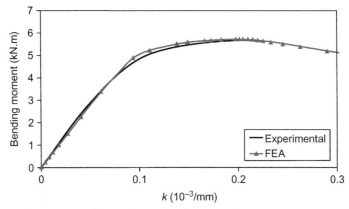

Figure 6.5 Comparison of experimental and numerical moment-curvature curves for specimen H-76 × 76 × 3 [6.5].

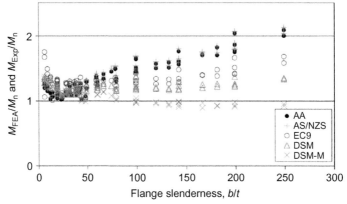

Figure 6.6 Comparison of experimental and numerical data with design strengths (M_n) [6.5].

with different geometries and aluminum alloy strengths. The nominal flexural strengths (unfactored design strengths) predicted by the American Specification [5.33] (M_{AA}), Australian/New Zealand Standard [5.34] ($M_{AS/NZS}$), and European Code [5.35] (M_{EC9}) for aluminum structures, as well as the current direct strength method [6.31] (M_{DSM}) and the modified direct strength method [6.32] (M_{DSM-M}) were compared with the bending strengths obtained from the parametric study (M_{FEA}) and experimental program (M_{Exp}) in the study [6.5]. Figure 6.6 shows a comparison of numerical (M_{FEA}) and experimental (M_{Exp}) results, which were nondimensionlized with respect to the design strengths (M_n) calculated using the American, Australian/New Zealand, and European Specifications for aluminum structures, as well as the current direct strength method and the modified direct strength method, against the nondimensional flange slenderness ratio (b/t) for aluminum beams. It was shown that the predictions given by the modified direct strength method were in best agreement with the numerical and test results. Hence, the modified direct strength method was recommended for the design of aluminum alloy flexural members of square hollow sections subjected to bending.

6.5. FINITE ELEMENT MODELING AND RESULTS OF EXAMPLE 3

The third example presented in this chapter is the simply supported cellular steel beams constructed from carbon steel modeled by Ellobody [6.17], shown in Figures 6.7 and 6.8. The full-scale destructive tests on simply supported cellular steel beams were conducted by Surtees and Liu [6.33], Warren [6.34], Tsavdaridis and D'Mello [6.35], and Tsavdaridis et al. [6.36]. The test specimens were denoted C1–C5 as given in Table 6.1. The definition of symbols for the cellular steel beams modeled by Ellobody [6.17] is shown in Figure 6.7. Specimen C1, tested in Ref. [6.33], had a length (L) of 5.25 m and was fabricated from hot-rolled I-section UB 406 × 140 × 39. The top and bottom flanges of C1 were braced laterally every 1 m as shown in Figure 6.8. The cellular steel beam

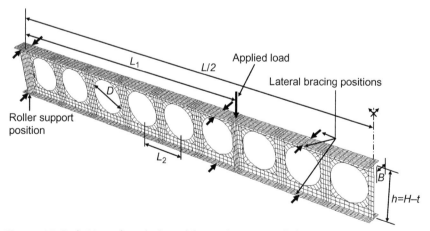

Figure 6.7 Definition of symbols and finite element mesh for the cellular steel beam C2 [6.17,6.34].

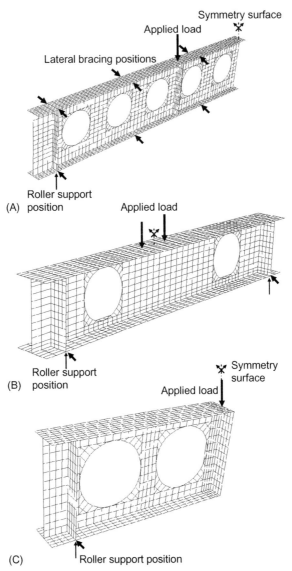

Figure 6.8 Finite element meshes for cellular steel beams modelled investigated [6.17,6.33−6.36]. (A) Specimen C1 tested in Ref. [6.33]. (B) Specimen C3 tested in Ref. [6.35]. (C) Specimens C4 and C5 tested in Ref. [6.36].

C1 was loaded with two concentrated loads, with a spacing from the support (L_1) as given in Table 6.1. The beam had a cell diameter (D) of 375 mm and a depth (H) of 581 mm, with D/H ratio of 0.65. The spacing between the centerlines of two adjacent circular cells (L_2) was

Table 6.1. Dimensions and Material Properties of Cellular Steel Beams Modeled in Ref. [6.17]

	Dimensions (mm)								Material Properties f_y (MPa)		Ref-erences
Test	H	B	t	s	L	L_1	L_2	h	D	Flange	Web	
C1	581.0	141.8	8.6	6.4	5250	1575	461.0	572.4	375	401.0	392.0	[6.33]
C2	463.2	101.6	7.0	5.8	7400	2466	400.0	452.5	325	401.0	392.0	[6.34]
C3	303.4	165.0	10.2	6.0	1500	600	900.0	293.2	231	337.5	299.0	[6.35]
C4	449.8	152.4	10.9	7.6	1700	850	409.5	438.9	315	359.6	375.3	[6.36]
C5	449.8	152.4	10.9	7.6	1700	850	378.0	438.9	315	359.6	375.3	[6.36]

461 mm, with L_2/D ratio of 1.23. The main failure mode observed experimentally for C1 was interaction of WDB and WPB failure modes (WDB + WPB). The cellular steel beam test specimen C2, tested in Ref. [6.34], had a length (L) of 7.4 m and was fabricated from hot-rolled I-section UB 305 × 102 × 25. The top and bottom flanges of C2 were braced laterally as shown in Figure 6.7. The cellular steel beam C2 was loaded with two concentrated loads. The cell diameter and depth of the beam were 325 and 463.2 mm, respectively, with D/H ratio of 0.7. The spacing between the centerlines of two adjacent circular cells was 400 mm, with L_2/D ratio of 1.23. Similar to C1, the main failure mode observed experimentally for C2 was interaction of WDB and WPB failure modes (WDB + WPB).

The cellular steel beam test C3, detailed in Ref. [6.35], had a length (L) of 1.5 m and was fabricated from hot-rolled I-section UB 305 × 165 × 40. The main purpose of the test was to investigate the shear resistance of cellular steel beams. The steel grade was S275. The cellular steel beam C3 was braced laterally as shown in Figure 6.8. The cell diameter and depth of the beam were 231 and 303.4 mm, respectively, with D/H ratio of 0.76. The cellular steel beam C3 had only two circular cells with spacing between the centerlines of the two cells given in Table 6.1. The main failure mode observed experimentally for C3 was interaction of LWB and SY failure modes (LWB + SY). Finally, the cellular steel beams C4 and C5, tested in Ref. [6.36], had a length (L) of 1.7 m and was fabricated from hot-rolled I-section UB 457 × 152 × 52. The steel grade was S355. The main variable parameter in the tests was the spacing between the web cells. The cellular steel beams C4 and C5 were braced laterally as shown in Figure 6.8. The circular cell diameter and depth of the beams were 315 and 449.8 mm, respectively, with D/H ratio of 0.7. The spacing between the centerlines of two adjacent circular cells were 409.5 and 378 mm, with L_2/D ratios of 1.3 and 1.2, respectively, for beams C4 and C5. Similar to C3, the main failure mode observed experimentally for C4 and C5 was interaction of LWB and SY failure modes (LWB + SY). Further details regarding the destructive full-scale cellular steel beam tests investigated in this study are given in Refs [6.33−6.36]. The test program provided useful and detailed data regarding the behavior of castellated steel beams. Therefore, finite element analysis was performed on the castellated steel beams.

The finite element program ABAQUS [1.27] was used in the analysis of cellular steel beams with circular holes tested by Surtees and Liu [6.33], Warren [6.34], Tsavdaridis and D'Mello [6.35], and Tsavdaridis et al. [6.36]. The models, developed by Ellobody [6.17], accounted for the measured geometry, initial geometric imperfections, and measured material properties of flange and web portions. A combination of 4-node and 3-node doubly curved shell elements with reduced integration S4R and S3R, respectively, were used to model the flanges and web of the cellular steel beams, as shown in Figures 6.7 and 6.8. Since lateral buckling of cellular steel beams is very sensitive to large strains, the S4R and S3R elements were used in this study to ensure the accuracy of the results. In order to choose the finite element mesh that provides accurate results with minimum computational time, convergence studies were conducted. It was found that approximately 40 × 50 mm (length by width of S4R element and depth by width of S3R element) ratio provides adequate accuracy in modeling the web while a mesh of approximately 35 × 50 mm was used in the flange.

The finite element analysis investigated in this study accounts for both geometrical and material nonlinearities. Hence, the application of boundary conditions is very important. Only half of the beams was modeled where there is exact symmetry in loading, geometry, and boundary conditions (see Section 3.3). The results obtained from modeling half of the beam has to be first compared with that obtained from modeling of the full beams and calibrated against the test results. In this study, it was found that modeling half of the beams C1, C2, C4, and C5 provide accurate results when verified against the test results while beam C3 was modeled in full as shown in Figures 6.7 and 6.8. Since the lateral bracing system used in the tests [6.33−6.36] was quite rigid preventing lateral transitional displacement and allowing rotational displacements, the top compression flange was prevented from lateral transitional displacement at the positions detailed in the tests. The load was applied in increments as static point load using the Riks method available in the ABAQUS [1.27] library. The stress—strain curve for the structural steel given in the EC3 [6.24] was adopted in Ref. [6.17] with measured values of the yield stress (f_y) and ultimate stress (f_u) used in the tests [6.33—6.36]. The material behavior provided by ABAQUS [1.27] (using the PLASTIC option) allows a non-linear stress—strain curve to be used (see Section 3.4).

Buckling of cellular beams depends on the lateral restraint conditions to compression flange and geometry of the beams. The lateral distortional buckling modes could be obtained by performing eigenvalue buckling analysis for cellular beams with actual geometry and actual lateral restraint conditions to the compression flange. Figures 6.9 and 6.10 show examples of unrestrained between ends and restrained buckling modes along the compression flange of cellular steel beams, respectively. Only the first buckling mode (eigenmode 1) is used in the eigenvalue analysis. Since buckling modes predicted by ABAQUS eigenvalue analysis [1.27] are generalized to 1.0, the buckling modes are factored by a magnitude of $L_u/1000$, where L_u is the length between points of effective bracing. The factored buckling mode is inserted into the load—displacement nonlinear analysis of the cellular beams following the eigenvalue prediction. It should be noted that the investigation of cellular steel beams with different slenderness ratios could result in LTB mode with or without WDB

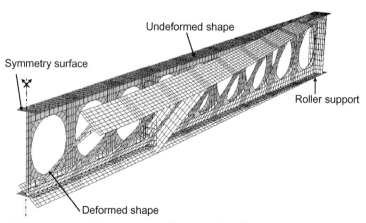

Figure 6.9 Unrestrained elastic lateral distortional buckling mode (eigenmode 1) for the cellular steel beam specimen C2 [6.17,6.34].

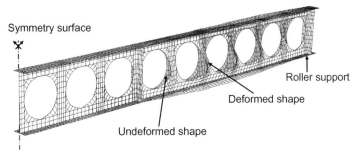

Figure 6.10 Restrained elastic lateral distortional buckling and WPB mode (eigen-mode 1) for the cellular steel beam specimen C5 [6.17,6.34].

and WPB modes. Hence, to ensure that the correct buckling mode is incorporated in the nonlinear displacement analysis, the eigenvalue buckling analysis must be performed for each cellular steel beam with actual geometry. More details regarding modeling of geometric imperfections can be found in Section 4.3. The cellular steel beams investigated in this study were fabricated by cutting a solid web of hot-rolled doubly symmetric I-section and reassembling it by shifting and welding the section components. Assuming that the cutting process is carefully conducted, the distribution of residual stresses within the cellular beam can be simulated as that of doubly symmetric I-sections. A typical distribution of residual stresses in hot-rolled doubly symmetric I-sections recommended in Ref. [6.37] was used in Ref. [6.17]. The residual stresses are implemented in the finite element model as initial stresses before applying loads. This can be performed using the (*INITIAL CONDITIONS, TYPE = STRESS) parameter available in the ABAQUS [1.27] library. Detailed information regarding modeling residual stresses can be found in Section 3.6.

The cellular steel beam tests detailed in Refs [6.33−6.36] were modeled by Ellobody [6.17]. The beams had different cross section geometries, lengths, steel strengths, and failure modes, which ensure that the model is capable to predict the inelastic behavior of a wide range of cellular steel beams. The failure loads, mid-span deflection at failure, failure modes, and load−mid-span deflection curves observed experimentally were compared with that predicted numerically using the finite element model. Table 6.2 gives a comparison between the failure loads (P_{Test}) and mid-span deflections at failure (δ_{Test}) obtained from the tests and finite element analyses (P_{FE}) and (δ_{FE}), respectively. It can be seen that good

Table 6.2 Comparison of Test and Finite Element Results of Cellular Steel Beams [6.17]

Test [Reference]	P_{Test} (kN)	δ_{Test} (mm)	Failure Mode	P_{FE} (kN)	δ_{FE} (mm)	Failure Mode	$\frac{P_{Test}}{P_{FE}}$	$\frac{\delta_{Test}}{\delta_{FE}}$
			Test			Finite Element Analysis		
C1 [33]	188.5	17.3	WDB + WPB	193.7	19.7	WDB + WPB	0.97	0.88
C2 [34]	114.0	48.0	WDB + WPB	113.0	51.2	WDB + WPB	1.01	0.94
C3 [35]	274.6	22.0	LWB + SY	281.9	20.3	LWB + SY	0.98	1.06
C4 [36]	288.7	19.6	WPB + SY	287.0	20.5	WPB + SY	1.01	0.96
C5 [36]	255.0	26.0	WPB + SY	263.9	28.0	WPB + SY	0.97	0.93
Mean	–	–	–	–	–	–	0.99	0.96
COV	–	–	–	–	–	–	0.020	0.080

agreement was achieved between the test and finite element results. The mean value of P_{Test}/P_{FE} and $\delta_{Test}/\delta_{FE}$ ratios are 0.99 and 0.96, respectively, with the COV of 0.02 and 0.08, respectively, as given in Table 6.2. Four failure modes were observed experimentally and confirmed numerically using the finite element model as summarized in Table 6.2. The cellular steel beams tested in Refs [6.33,6.34] underwent WDB that followed by WPB. The cellular steel beam tested in Ref. [6.35], having a length of 1.5 m, underwent LWB followed by SY. The SY failure mode was predicted from the finite element model by comparing the von Mises stresses in the cellular steel beams at failure against the measured yield stresses. On the other hand, the SY was judged in the tests by comparing the test failure loads against the plastic collapse loads (P_{px}) calculated according to AS4100 [6.16]. Finally, the cellular steel beams tested in Ref. [6.36], having a length of 1.7 m, underwent WPB followed by SY. The load−mid-span deflection curves predicted experimentally and numerically were also compared as shown in Figures 6.11 and 6.12 as examples for beam tests C1 and C2. It can be shown that generally good agreement was achieved between experimental and numerical relationships.

Furthermore, the deformed shapes of cellular steel beams at failure observed experimentally and numerically were compared. Figure 6.13 shows an example of the displaced shape observed in the test specimen C3 in comparison with that predicted from the finite element analysis. It can be seen that the experimental and numerical deformed shapes are in good agreement. The failure mode observed experimentally and confirmed numerically was a combination of LWB and SY. The data obtained from

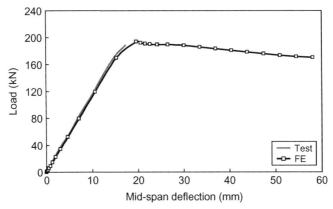

Figure 6.11 Comparison of load–mid-span deflection curves for test specimen C1.

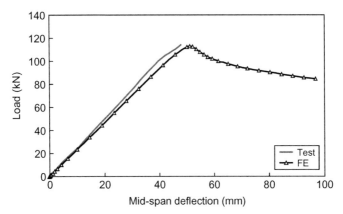

Figure 6.12 Comparison of load–mid-span deflection curves for test specimen C2.

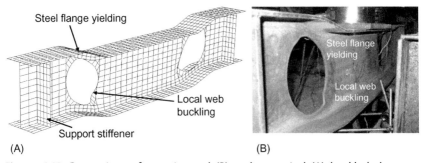

Figure 6.13 Comparison of experimental (B) and numerical (A) buckled shapes at failure for Specimen C3 [6.17,6.35].

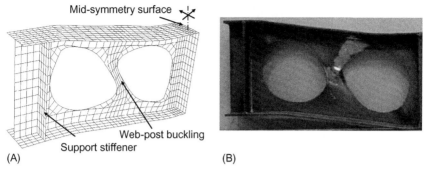

Mid-symmetry surface

Web-post buckling
Support stiffener

(A)   (B)

Figure 6.14 Comparison of experimental (B) and numerical (A) buckled shapes at failure for Specimen C4 [6.17,6.36].

ABAQUS [1.27] have shown that the von Mises stresses at the maximum stressed fibers at the top and bottom flanges under the applied load exceeded the measured yield stresses. Similarly, Figure 6.14 shows another example of the displaced shape observed in the test specimen C4 in comparison with that predicted from the finite element analysis. Once again, it can be seen that good agreement exists between the experimental and numerical deformed shapes. The failure mode observed experimentally and confirmed numerically was a combination of WPB and SY.

The verified finite element model developed in Ref. [6.17] was used to study the effects of the change in cross section geometries, beam length, steel strength, and nondimensional slenderness on the strength and buckling behavior of cellular steel beams. One hundred and twenty cellular steel beams were analyzed using the finite element model. The beams were divided into 12 groups denoted G1−G12. The first six groups G1−G6 had beams with a length of (L) 5250 mm and a depth (H) of 581 mm, while groups G7−G12 had beams with a length of 7400 mm and a depth of 463.2 mm, which is similar to the beam lengths of test specimens C1 and C2, respectively. The cellular steel beams in G1−G6 had a cell diameter (D) of 325 mm and spacing between centerlines of two adjacent cells (L_2) of 525 mm, with (D/H) and (L_2/D) ratios of 0.56 and 1.62, respectively. On the other hand, the cellular steel beams in G7−G12 had the same D of 325 mm and L_2 of 400 mm, with (D/H) and (L_2/D) ratios of 0.7 and 1.23, respectively. Group G1 had 10 cellular steel beams S1−S10 having a height (h) of 572.4, a width (B) of 141.8 mm, and a web thickness (s) of 6.4 mm, but with different flange thickness (t) ranging from 4 to 16 mm. This has resulted in B/t ratios ranging from 35.5 to 8.9. Group G1 had a steel yield stress (f_y) of 275 MPa and an ultimate stress (f_u) of 430 MPa. Groups G2 and G3 were identical to G1 except with f_y of 460, and 690 MPa and f_u of 530 and 760 MPa, respectively. The yield and ultimate stresses conform to EC3 [6.24]. Group G4 had 10 specimens S31−S40 having h of 572.4, B of 141.8 mm, and t of 8.6 mm but with different s ranging from 4 to16 mm. This has resulted in h/s ratios varying from 143.1 to 35.8. Groups G5 and G6 were identical to G4 but with different steel yield and ultimate stresses. Groups G4−G6 had the same steel stresses as G1−G3, respectively.

Group G7 had 10 cellular steel beams S61−S70 having h of 452.5, B of 123.3 mm, and s of 7.1 mm, but with different t ranging from 4 to 16 mm. This has resulted in B/t ratios ranging from 30.8 to 7.7. Group G7 had a steel yield stress (f_y) of 275 and an ultimate stress (f_u) of 430 MPa. Groups G8 and G9 were identical to G7 except with f_y of 460 and 690 MPa and f_u of 530 and 760 MPa, respectively. Group G10 had 10 specimens S91−S100 having h of 452.5, B of 123.3 mm, and t of 10.7 mm but with different s ranging from 4 to16 mm. This has resulted in h/s ratios varying from 113.1 to 28.3. Groups G11 and G12 were identical to G4 but with different steel yield and ultimate stresses. Groups G10−G12 had the same steel stresses as G7−G9, respectively. The investigated cellular steel beams had different nondimensional slenderness (λ) calculated based on AS4100 [6.16] ranged from 0.66 to 1.93. The nondimensional slenderness (λ) is equal to the square root of the major axis full plastic moment divided by the elastic buckling moment, and is considered as a guide for beam slenderness in this study.

To date, there is no design guides in current codes of practice that account for the inelastic behavior of normal and high strength cellular steel beams under combined buckling modes including WDB. Only design guides were found in the AS4100 [6.16] that considers LTB of doubly symmetric I-sections as well as design guides in the AISC [6.38] that controls the errors associated with neglecting web distortion in doubly symmetric I-sections. Zirakian and Showkati [6.27] have concluded that the AISC [6.38] predictions are overconservative and in some cases may cause economic losses for doubly symmetric I-sections under distortional buckling. In the study [6.17], the failure loads of the cellular steel beams investigated in the parametric study were compared with the design guides given in the AS4100 [6.16]. Following the AS4100 design guides [6.16], the nominal buckling moment strength (M_b) of compact doubly symmetric I-section beams is given by

$$M_b = \alpha_m \alpha_s M_{px} \qquad (6.1)$$

where $M_{px} = f_y S_x$ is the major axis full plastic moment corresponding to collapse load P_{px}, f_y is the yield stress, S_x is the plastic section modulus, α_m is a moment modification factor which allows for nonuniform moment distributions (taken 1.0 for simply supported beams under two concentrated loads), and α_s is a slenderness reduction factor which allows for the effects of elastic buckling, initial geometric imperfections, initial twist, and residual stresses, and which is given by Trahair [6.39] as follows:

$$\alpha_s = 0.6 \left(\sqrt{\left(\frac{M_{px}}{M_{yz}}\right)^2 + 3} - \left(\frac{M_{px}}{M_{yz}}\right) \right) \le 1.0 \qquad (6.2)$$

where M_{yz} is the elastic buckling moment of a simply supported beam in uniform bending given by

$$M_{yz} = \sqrt{\frac{\pi^2 E I_y}{L_u^2} \left(GJ + \frac{\pi^2 E I_w}{L_u^2} \right)} \qquad (6.3)$$

where E and G are the Young's modulus and shear modulus of elasticity, I_y, J and I_w are the minor axis section moment of area, the uniform torsion section constant, and the warping section constant, respectively. The nondimensional slenderness (λ) of the cellular steel beam according to AS4100 is equal to $\sqrt{M_{px}/M_{yz}}$ and is considered as a guide for beam slenderness in this study. The design load of cellular steel beams, with simply supported ends under two concentrated loads, based on AS4100 (P_{AS4100}) is calculated from M_b.

Figure 6.15 plots the failure loads of cellular steel beams in groups G1−G3 predicted from the finite element analysis (P_{FE}) and design guides (P_{AS4100}). The failure loads were plotted, as a percentage of the plastic collapse load (P_{px}), against the nondimensional flange width-to-thickness ratio (B/t). Looking at the cellular steel beams in G1, it can be seen that as the B/t ratio increased from 8.9 to 20.3, the P_{FE}/P_{px} ratio is increased nonlinearly. While increasing the B/t ratio above 20.3 has resulted in approximately nonlinear decrease in the P_{FE}/P_{px} ratio. However, interestingly, as the B/t ratio increased from 8.9 to 35.5, the P_{AS4100}/P_{px} ratio is reduced in a nonlinear relationship. This is attributed to the fact that for cellular steel beams having less B/t, the failure mode was dominated by the presence of combined (WDB + WPB), which was not considered by the specification [6.16]. The comparison has also shown that the AS4100

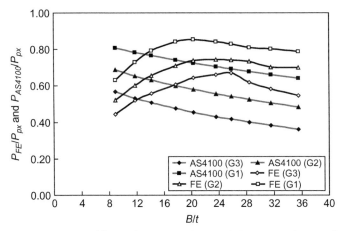

Figure 6.15 Comparison of finite element analysis and design predictions for cellular steel beams in groups G1–G3 [6.17].

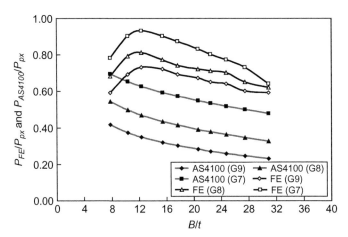

Figure 6.16 Comparison of finite element analysis and design predictions for cellular steel beams in groups G7–G9 [6.17].

design guides are generally conservative for the cellular steel beams with normal yield strength (beams in G1), except for beams S9 and S10 failing by (LTB + WDB + WPB). Similar conclusions were observed for beams in G2 and G3 having higher steel strengths with f_y of 460 and 690 MPa, respectively, as shown in Figure 6.15. However, it can be seen that the specification predictions were quite conservative for the beams with higher yield stresses (beams in G2 and G3) and failing by LTB. Similar

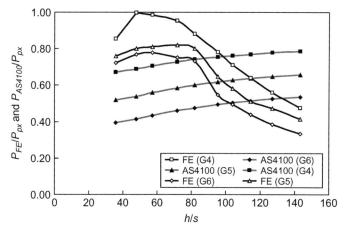

Figure 6.17 Comparison of finite element analysis and design predictions for cellular steel beams in groups G4−G6 [6.17].

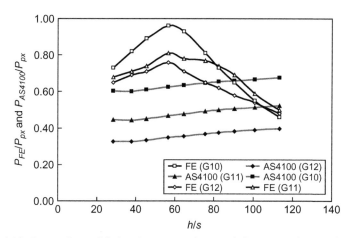

Figure 6.18 Comparison of finite element analysis and design predictions for cellular steel beams in groups G10−G12 [6.17].

conclusions could be drawn for the cellular steel beams in G7−G9 as shown in Figure 6.16.

Figures 6.17 and 6.18 plots the P_{FE}/P_{px} and P_{AS4100}/P_{px} ratios against the nondimensional web height-to-web thickness ratio (h/s) for the cellular steel beams in G4−G6 and G10−G12. Looking at Figure 6.17, it can be seen that there is a dramatic decrease in the failure load of cellular steel beams having h/s ratios greater than or equal to 104.1 and failing mainly owing to the combined WDB and WPB failure mode. It can also be seen that some cellular steel beams with f_y of 275 MPa failed due to plastic collapse, while none of the cellular steel beams with f_y of 460 and 690 MPa exceeded the plastic resistance. The specification predictions were unconservative for the cellular steel beams undergoing combined WDB and WPB failure mode. The specification predictions were generally conservative for cellular steel beams with normal yield strength and failing mainly by LTB. However, the specification predictions were quite conservative for cellular steel beams with higher yield stresses and once again failing mainly by LTB. Similar conclusions could be drawn for the cellular steel beams in G10–G12 as shown in Figure 6.18.

REFERENCES

[6.1]   Liu, T. C. H. and Chung, K. F. Steel beams with large web openings of various shapes and sizes: Finite element investigation. Journal of Constructional Steel Research, 59(9), 1159–1176, 2003.

[6.2]   Chung, K. F., Liu, T. C. H. and Ko, A. C. H. Investigation on Vierendeel mechanism in steel beams with circular web openings. Journal of Constructional Steel Research, 57(5), 467–490, 2001.

[6.3]   Redwood, R. G. and McCutcheon, J. O. Beam tests with un-reinforced web openings. Journal of Structural Division, Proceedings of ASCE, 94(ST1), 1–17, 1968.

[6.4]   Mohri, F., Brouki, A. and Roth, J. C. Theoretical and numerical stability analyses of unrestrained, mono-symmetric thin-walled beams. Journal of Constructional Steel Research, 59(1), 63–90, 2003.

[6.5]   Zhu, J. H. and Young, B. Design of aluminum alloy flexural members using direct strength method. Journal of Structural Engineering, ASCE, 135(5), 558–566, 2009.

[6.6]   Liu, Y. and Gannon, L. Finite element study of steel beams reinforced while under load. Engineering Structures, 31(11), 2630–2642, 2009.

[6.7]   Gannon, L. Strength and behavior of beams strengthened while under load. MSc thesis. Halifax (NS, Canada): Department of Civil and Resource Engineering, Dalhousie University; 2007.

[6.8]   Wu, Z. Q. and Grondin, G. Y. Behavior of steel columns reinforced with welded steel plates. Structural Engineering Report no. 250. Edmonton (AB, Canada): Department of Civil and Environmental Engineering, University of Albert; 2002.

[6.9]   Theofanous, M. and Gardner, L. Experimental and numerical studies of lean duplex stainless steel beams. Journal of Constructional Steel Research, 66(6), 816–825, 2010.

[6.10] Theofanous, M., Chan, T. M. and Gardner, L. Structural response of stainless steel oval hollow section compression members. Engineering Structures, 31(4), 922–934, 2009.

[6.11] Theofanous, M., Chan, T. M. and Gardner, L. Flexural behaviour of stainless steel oval hollow sections. Thin-Walled Structures, 47(6–7), 776–787, 2009.

[6.12] Sweedan, A. M. I. Elastic lateral stability of I-shaped cellular steel beams. Journal of Constructional Steel Research, 67(2), 151–163, 2011.

[6.13] Haidarali, M. R. and Nethercot, D. A. Finite element modelling of cold-formed steel beams under local buckling or combined local/distortional buckling. Thin-Walled Structures, 49(12), 1554−1562, 2011.

[6.14] CUFSM Version 3.12, by Ben Schafer. Department of Civil Engineering, Johns Hopkins University; 2006.

[6.15] Ellobody, E. Interaction of buckling modes in castellated steel beams. Journal of Constructional Steel Research, 67(5), 814−825, 2011.

[6.16] Australian Standards AS4100. Steel Structures. AS4100-1998. Sydney, Australia: Standards Australia, 1998.

[6.17] Ellobody, E. Nonlinear analysis of cellular steel beams under combined buckling modes. Thin-Walled Structures, 52(3), 66−79, 2012.

[6.18] Anapayan, T. and Mahendran, M. Numerical modelling and design of LiteSteel Beams subject to lateral buckling. Journal of Constructional Steel Research, 70, 51−64, 2012.

[6.19] Kurniawan, C. W. and Mahendran, M. Elastic lateral buckling of simply supported LiteSteel Beams subject to transverse loading. Thin-Walled Structures, 47(1), 109−119, 2009.

[6.20] Zhou, F., Young, B. and Lam, H. C. Welded aluminum alloy plate girders subjected to shear force. Advanced Steel Construction, 8(1), 71−94, 2012.

[6.21] Soltani, M. R., Bouchaïr, A. and Mimoune, M. Nonlinear FE analysis of the ultimate behavior of steel castellated beams. Journal of Constructional Steel Research, 70, 101−114, 2012.

[6.22] Lusas 13. LUSAS: finite element stress analysis system: user's manual. Finite Element Analysis Ltd. Version 13.5, UK, 2003.

[6.23] Kankanamge, N. D. and Mahendran, M. Behaviour and design of cold-formed steel beams subject to lateral-torsional buckling. Thin-Walled Structures, 51, 25−38, 2012.

[6.24] European Committee for Standardization (ECS), Eurocode 3. Design of steel structures—Part 1.3, general rules—supplementary rules for cold-formed thin gauge members and sheathing. Brussels: Eurocode 3, 2006.

[6.25] Standards Australia/New Zealand. Cold-formed steel structures. AS/NZS 4600. Sydney, Australia: 2005.

[6.26] American Iron and Steel Institute (AISI). North American specification for the design of cold-formed steel structural members. Washington DC: American Iron and Steel Institute, 2007.

[6.27] Zirakian, T. and Showkati, H. Distortional buckling of castellated beams. Journal of Constructional Steel Research, 62(9), 863−871, 2006.

[6.28] Bradford, M. A. Lateral-distortional buckling of steel I-section members. Journal of Constructional Steel Research, 23(1-3), 97−116, 1992.

[6.29] Nethercot, D. A. and Trahair, N. S. Inelastic lateral buckling of determinate beams. Journal of Structural Division, ASCE, 102(ST4), 701−717, 1976.

[6.30] Zhu, J. H. and Young, B. Experimental investigation of aluminum alloy thin-walled tubular members in combined compression and bending. Journal of Structural Engineering, ASCE, 132(12), 1955−1966, 2006.

[6.31] Schafer, B. W. and Peköz, T. Direct strength prediction of cold-formed steel members using numerical elastic buckling solutions. Proceedings of 14th International Specialty Conference on Cold-Formed Steel Structures, University of Missouri-Rolla, Rolla, MO; 69−76, 1998.

[6.32] Zhu, J. H. and Young, B. Aluminum alloy tubular columns—Part II: Parametric study and design using direct strength method. Thin-Walled Structures, 44(9), 969−985, 2006.

[6.33] Surtees, J. O. and Liu, Z. Report of loading tests on cellform beams. Research report, University of Leeds; 1995.

[6.34] Warren, J. Ultimate load and deflection behaviour of cellular beams. MSc thesis. Durban: School of Civil Engineering, University of Natal; 2001.

[6.35] Tsavdaridis, K. D. and D'Mello, C. Web buckling study of the behaviour and strength of perforated steel beams with different novel web opening shapes. Journal of Constructional Steel Research, 67, 1605–1620, 2011.

[6.36] Tsavdaridis, K. D., D'Mello, C., and Hawes, M. Experimental study of ultra shallow floor beams with perforated steel sections. Research report, National Specialist Contractors' Council, NSCC2009; 312–319, 2009.

[6.37] Chen, S. and Jia, Y. Numerical investigation of inelastic buckling of steel-concrete composite beams prestressed with external tendons. Thin-Walled Structures, 48, 233–242, 2010.

[6.38] AISC. Specification for structural steel buildings. American Institute for Steel Construction, Reston, Chicago, IL, ANSI/AISC 360-05 2005.

[6.39] Trahair, N. S. Flexural-torsional buckling of structures. London: E & FN Spon, 1993.

CHAPTER 7

Examples of Finite Element Models of Metal Tubular Connections

7.1. GENERAL REMARKS

The finite element modeling of metal columns and beams have been highlighted in Chapters 5 and 6, respectively, and examples of finite element models of metal connections are presented in this chapter. The connections investigated in this book can be constructed from any metal material. The connections investigated can be different boundary conditions at the ends and can be rigid or semi-rigid connections. In addition, the connections investigated can have different cross sections constructed from hot-rolled, cold-formed, and welded sections. It should be noted that there are many types of connections that are used in practice, which may need a separate book to provide full details. However, this book provides an approach for modeling metal connections which can be applied to different types of connections. Therefore, based on recent investigations by the authors, it is decided to present an innovative type of metal connections which are cold-formed stainless steel tubular joints. Cold-formed stainless steel tubular joints are being used increasingly for architectural and structural purposes due to their aesthetic appearance, high corrosion resistance, ductility property, improved fire resistance, and ease of construction and maintenance. The practical applications of cold-formed welded tubular joints are shown in Figures 7.1–7.3. To date, there is little research being carried out on cold-formed stainless steel tubular joints. Furthermore, the current design rules for stainless steel tubular joints are mainly based on the carbon steel sections. The mechanical properties of stainless steel sections are clearly different from those of carbon steel sections. Stainless steel sections have a rounded stress–strain curve with no yield plateau and low proportional limit stress compared to carbon steel sections, as shown in Figure 1.1. To facilitate the use of stainless steel tubular structures, design guidelines should be prepared for

Finite Element Analysis and Design of Metal Structures
DOI: http://dx.doi.org/10.1016/B978-0-12-416561-8.00007-X

Figure 7.1 Curtain wall construction of Langham Place in Hong Kong, China.

Figure 7.2 Roof structure of Kuala Lumpur International Airport, Malaysia.

stainless steel tubular hollow sections to offset its high material costs through efficient design.

Numerical investigations on cold-formed stainless steel tubular T-joints, X-joints, and X-joints with chord preload are current research

Figure 7.3 Footbridge in Singapore.

topics. The stainless steel tubular joints, highlighted in this book, were fabricated from square and rectangular hollow sections (SHS and RHS) brace and chord members. A test program carried out on a wide range of cold-formed stainless steel tubular T- and X-joints of SHS and RHS is introduced in this chapter. The corresponding finite element models developed based on the experimental investigations are described in detail. This chapter highlights how the geometric and material nonlinearities of stainless steel tubular joints can be carefully incorporated in the finite element models. This chapter also presents the results obtained from experimental investigations and finite element analyses comprising the joint strengths, failure modes, and load−deformation curves of stainless steel tubular joints.

This chapter initially presents a survey of recent published numerical, using finite element method, investigations on metal connections. After that, the chapter presents the experimental investigations and finite element models previously published by the authors for different metal tubular connections. The joint strengths, failure modes, and load−deformation curves of cold-formed stainless steel tubular T- and X-joints obtained experimentally and numerically are highlighted and discussed to show the effectiveness of the results. In addition, this chapter presents the comparison between the experimental and finite element analysis results to calibrate the developed finite element models. The authors highlight how the information presented in the previous chapters is used to develop the examples of finite element models to accurately simulate the structural behavior of test specimens introduced in this chapter. The authors have an aim that the examples highlighted in this chapter can explain to readers the effectiveness of finite element models in providing detailed data that augment experimental investigations conducted in the field. The results are discussed to show the significance of the finite element models in predicting the structural response of different metal structural connections. The authors also have an aim that by highlighting the structural performance of metal tubular connections, researchers can use the same approach to investigate the building structural behavior.

7.2. PREVIOUS WORK

Finite element analysis can be used effectively to highlight the performance of metal connections, with many different general-purpose finite element analysis software can be used such as ANSYS [5.3], ABAQUS [1.27], I-DEAS [7.1], MARC [7.2], PAFEC [7.3], and PATRAN [7.4]. Finite element analysis can be used to investigate the behavior of cold-formed welded circular hollow section (CHS), SHS, and RHS tubular joints, which are discussed in this chapter. The accuracy of finite element analysis of metal connections mainly depends on the use of proper finite element type, material modeling, analysis procedure, integration scheme, loading and boundary condition, mesh refinement, and modeling of weld shape, as presented in the previous chapters.

Extensive investigations were presented in the literature detailing the performance of metal connections through using the finite element analysis. Packer [7.5] undertook a parametric study to identify the principal factors that affect the behavior and ultimate strength of statically loaded welded joints in RHS steel trusses, having one compression bracing member and one tension bracing member. The numerical investigation was validated by comparing with a large number of test results, which mainly examined cases in which the bracing members are gapped as well as overlapped at the chord face connection. The influential factors investigated in the parametric study include the yield stress, chord force, strut dimensions relative to the chord in overlapped joints, amount of gap or overlap, orientation of bracing members in gapped joints, and width ratio between bracing members and chord in gapped joints and the bracing member angles. Simplified design formulae were also proposed to predict the ultimate joint strength.

Packer et al. [7.6] described a finite element model for a welded gapped K-joint in a RHS warren truss to study the parameters influencing the flexibility of joint at the serviceability limit state. A bilinear elastic-strain hardening material characteristic was incorporated and large deflection behavior of both the chord face and the supporting frames was included in the model. It was found that optimum correlation between chord face deformations is primarily dependent on the stress distribution assumed around the perimeter of the branch members at the chord face junction, with the flexibility of the inclusions beneath the branch members also being influential. Ebecken et al. [7.7] established nonlinear elasto-plastic finite element models to determine the static strength of typical X-joint under axial brace loads. Techniques for automatically generating finite element meshes in stress analysis of tubular intersections were used. Only one-eighth of the tubular joint was modeled due to the symmetric loading and boundary conditions. The results were obtained by using 3-node flat shell element and 8-node isoparametric shell element. The modeling and computational aspects which are required for dealing with the elasto-plastic analysis were discussed to obtain the ultimate joint strengths.

Zhang et al. [7.8] developed a suitable nonlinear finite element model that incorporated large deflection to conduct elasto-plastic analysis of the ultimate strength of welded RHS joints. The comparison between experimental and numerical results showed good agreement. A model called the equivalent frame tube model which can properly simulate the characteristic of the ultimate strength of RHS joints was proposed. A design formula was derived to predict the ultimate strength of RHS X-joints. Moffat et al. [7.9] carried out nonlinear finite element analysis to assess the static collapse strength of a sample tubular T-joint configuration subjected to compressive brace loading. Two series of models were used to assess the effects of varying the chord length and chord boundary conditions on the ultimate joint strengths. The finite element model was produced using the PATRAN [7.4] mesh generation program. The static strengths of the various models were determined using the ABAQUS [1.27] finite element program. Three-dimensional brick elements (C3D20 and C3D15) were employed with elastic−perfectly plastic material properties being used in the finite element analysis. It was shown that chord length and boundary conditions can have a significant influence on static collapse loads.

Saidani [7.10] investigated the effect of joint eccentricity on the local and overall behavior of truss girders made from RHS members. Three typical truss girders with identical general layout and comprising different joint eccentricities were analyzed. Different numerical models of analysis were presented and the implications for design were discussed. Only half of each truss was analyzed due to the symmetry in loading and boundary conditions. It was shown that the connection eccentricity can have significant effects on the axial force distribution in the bracings. However, its influence on the overall truss deflection was negligible. Lee [7.11] reviewed the numerical modeling techniques used in the finite element analysis of tubular joints and provided guidance on obtaining information on strength, stress fields, and stress intensity factors. Several commercial software packages were compared in the mesh generation of complex intersections of tubular member. For strength analysis, guidance was given on model discretization, choice of elements, material properties input, and weld modeling for valid results and modeling limitations. For stress analysis, guidance was given on the extraction of stresses by using different types of elements, weld modeling, and the use of submodeling techniques for fatigue calculations. For fracture analysis, guidance was given on the use of line-spring elements in shell models, the choice of solid elements, cracked mesh generation, and interpretation of stress intensity factors from finite element outputs.

Choo et al. [7.12] presented a new approach in the definition of joint strength for thick-walled CHS X-joint subjected to brace axial loading. The finite element models were created using MSC/PATRAN [7.4], and the nonlinear analysis incorporating geometric and material nonlinearities was carried out using the general-purpose finite element software ABAQUS [1.27]. Nonlinear material property for the numerical study was based on the true stress—strain curve represented by piecewise linear relations. Twenty- quadratic solid elements with reduced integration (C3D20R) given in the ABAQUS [1.27] were consistently used for modeling the brace, chord, and weld regions. For thick-walled joints, four layers of elements were employed across the chord and brace wall thickness, while for thin-walled joints, two layers of elements were specified. For X-joint under brace axial load, one-eighth of the entire joint was simulated with proper boundary conditions applied on the symmetry planes. A new approach in the definition of joint strength based on the plastic load approach with a consistent coefficient λ was found to provide consistent correlation with the peak loads in the joint load—deformation curve. The joint strength predicted using the present approach was also compared with that obtained from the current design recommendations for thick-walled CHS X-joint.

Karamanos and Anagnostou [7.13] presented nonlinear finite element model to investigate the influence of external hydrostatic pressure on the ultimate capacity of uniplanar X- and T-welded tubular connections under axial and bending loads. The general-purpose finite element program ABAQUS [1.27] was employed for the simulation and the nonlinear analysis of tubular joint. A nonstructured mesh of finite elements was constructed, using a 20-node three-dimensional solid element with reduced integration (C3D20R with $2 \times 2 \times 2$ Gauss point grid). Only one-eighth of the tubular X-joint was modeled with appropriate symmetry conditions on the symmetry edges. The finite element meshes are quite fine in the vicinity of the weld profile due to the strain concentrations and the occurrence of plastic deformations, but are rather coarse away from the weld. Two elements were used through the tube thickness to accurately simulate tube wall bending. Both geometric and material nonlinearities were considered in the finite element analysis. Inelastic effects on the response were taken into account through von Mises large strain plasticity model with isotropic hardening. The nonlinear analysis was conducted using a displacement-controlled marching scheme to trace unstable equilibrium paths that exhibit "snap-back." Good agreement between the numerical results and test data was found. It was shown that external pressure causes structural instability and has significant effects on both the ultimate load and the deformation capacity of the joints. A simple analytical formulation was developed to yield closed-form expressions for the load—deformation relationship, which approximated the elastic—plastic response of a pressurized tubular X-joint under axial loads.

Gho et al. [7.14] presented the experimental and numerical results of the ultimate load behavior of CHS tubular joints with complete overlap of braces. The finite element package MARC [7.2] with pre- and post-processing program MENTAT was adopted for the simulation. The mid-surface of the wall thickness of joint members was modeled using doubly curved 4-node thick shell elements (MARC element type 75), which can be used to consider the effect of transverse shear deformation. Only one-half of the joint was modeled due to the symmetrical geometry and boundary conditions. Fine meshes were used at the intersections of members and the gap region of the joint to account for the effect of high stress gradients. Both material and geometric nonlinearities were included in the finite element analysis. The von Mises yield criterion and the multilinear isotropic work hardening rule of plasticity were applied. The large displacement, the updated Lagrange procedure, and the finite strain plasticity were activated in the analysis for complete large strain plasticity formulation. A full Newton—Raphson method was adopted to reassemble the stiffness matrix at each of the iteration. A modified Riks—Ramm method was adopted in the loading application. A mesh convergence study was performed to obtain optimum mesh size. A detailed parametric study including 1296 FE models was conducted by using the verified finite element model to examine the failure modes and the load—deformation characteristics of the joint. There were four possible failure modes of the joint under lap brace axial compression. A combination of these failure modes can occur depending on the geometrical parameters of the joints.

Van Der Vegte et al. [7.15] conducted extensive numerical research into the chord stress effect of CHS uniplanar K-, T-, and X-joints. The study presented the results of finite element analyses on CHS uniplanar X-joints under axial brace load with the chord subjected to either axial load, in-plane bending moment or combinations of axial load and in-plane bending moment. A new strength formulation was established for X-joints under axial brace load solely. A chord stress function was derived, describing the combined effect of axial chord load and in-plane bending chord moment on the ultimate strength of uniplanar X-joints.

Gho and Yang [7.16] presented both experimental and numerical investigations on CHS tubular joint with complete overlap of braces. A doubly curved thick shell element (Type 75) given in the finite element package MARC [7.2] was used to model the midface of members of the joint, which took the transverse shear deformation effect into consideration. The weld was also carefully modeled as a ring of shell elements around the joint intersection. Only one-half of the joint was modeled in view of the symmetrical properties of geometry and boundary conditions. The fine meshes were created at the joint intersections and the short segment of through brace to account for the effect of high stress gradients. The convergence study by varying the mesh density at joint intersections was also conducted to obtain the optimum mesh size. The nonlinearity of material and geometrical properties of the joint was included in the finite element analysis. The von Mises yield criterion and the multilinear isotropic work hardening rule of plasticity were applied. The large displacement, the updated Lagrange procedure, and the finite strain plasticity were all activated for complete large strain plasticity formulation. A full Newton—Raphson method was adopted to reassemble the stiffness matrix at each of iteration. A modified Riks—Ramm method was employed in the loading application. The finite element model was verified against the current and previous test results with good agreement. A parametric equation for the prediction of ultimate joint strength was developed based on 3888 finite element models in the parametric study. It was found that the ultimate joint strength was not significantly affected by the boundary conditions and the chord prestresses. However, the ultimate joint strength decreased with increasing gap size.

Shao et al. [7.17] studied the effect of the chord reinforcement on the static strength of welded tubular T-joints by using the finite element method. Twenty-Node hexahedral solid elements were used to model brace and chord members as well as the butt weld connecting the tubes with different thicknesses. The mesh around the weld toe was refined, while a relatively coarse mesh was used far away from the weld toe to increase the computational efficiency and obtain accurate numerical results. Six layers of elements were employed in the thickness direction of the chord reinforced region around the weld toe to consider the high stress concentration in this region, while two layers of elements were used in the regions far away from the weld toe to save computational time. The residual stresses in the heat-affected region of the T-joint were not considered in the finite element analysis. It was found that the static strength can be greatly improved by increasing the chord thickness near the intersection. However, it is ineffective to improve the static strength by increasing the length of the reinforced chord. Furthermore, a parametric study of 240 T-joints was performed to investigate the effects of the geometrical parameters and the chord thickness on the static strength of the T-joints. A parametric equation was proposed to predict the static strength of the tubular T-joint subjected to axial compression.

Van der Vegte et al. [7.18] reviewed the nonlinear finite element analyses in the field of tubular structures. The main aspects of finite element analyses for welded hollow section joints were overall discussed, which include the solution technique such as implicit versus explicit methods, choice of element type, material nonlinearity, modeling of the welds, and limitations in predicting certain failure modes. The main difference between the implicit and explicit solvers was further highlighted in terms of the solution strategy and application as well as the effect of mesh refinement on computational time and memory requirements, which was also illustrated in the given examples. Liu and Deng [7.19] studied the effect of out-of-plane bending performance of CHS X-joints by using finite element analysis. Three-node triangular shell elements with reduced integration (S3R) given in the ABAQUS [1.27] were used to model the center region, while 4-node shell elements with reduced integration (S4R) were used to model the other regions and the end plates of the chord. The Riks method was used to consider effect of geometrically nonlinear performance while the ideal elasto-plastic constitutive model was employed. The finite element model was verified to be credible by comparing with the experimental results. The influence of brace inclination angles, axial stress and diameter-to-thickness ratio of chord on failure mode, ultimate bearing capacity, and flexural rigidity of tubular joints were all investigated in the parametric study. It was found that a larger brace inclination angle could increase the ultimate capacity of X-joint but could decrease its flexural rigidity. Furthermore, axial compression and tension in chords could weaken the rigidity and bearing capacity of X-joints, and a considerable axial compression could lower the ductility of X-joints.

7.3. EXPERIMENTAL INVESTIGATIONS OF METAL TUBULAR CONNECTIONS

7.3.1 General

The experimental investigation, presented in this chapter and used for the verification of the following finite element models, was conducted by Feng and Young [7.20,7.21] on cold-formed stainless steel tubular T- and X-joints. The test specimens were fabricated from SHS and RHS brace and chord members. Both high strength stainless steel (duplex and high strength austenitic) and normal strength stainless steel (AISI 304) specimens were tested. Special attention was given to the deformations of stainless steel tubular joints, which were generally larger than those of carbon steel tubular joints. The test strengths, flange indentation, and web deflection of chord members, as well as the observed failure modes for all test specimens were obtained. The tests were well designed and instrumented, which agrees with the criteria of a successful test programs set in Section 1.3.

7.3.2 Scope

The design of the test program has accounted for most of the parameters affecting the behavior of tubular joints. Looking at the strength of stainless steel tubular joints, which depends mainly on (i) the ratio of brace width to chord width ($\beta = b_1/b_0$), (ii) the ratio of brace thickness to chord thickness ($\tau = t_1/t_0$), (iii) the ratio of chord width to chord thickness ($2\gamma = b_0/t_0$), and (iv) the compressive preload (N_p) applied to the chord members. Therefore, the tests were conducted by applying axial compression force to the brace members using different values of β ranged from 0.5 to 1.0 (full width joint), τ from 0.5 to 2.0, and 2γ from 10 to 50, which is beyond the validity range of most current design specifications for tubular connections ($2\gamma \leq 35$). Three different levels of compressive preload were applied to the chord members of the stainless steel tubular X-joints. The effect of compressive chord preload on the strength of stainless steel tubular X-joint was evaluated.

7.3.3 Test Specimens

The compression tests were performed on cold-formed stainless steel tubular T- and X-joints of SHS and RHS. A total of 22 stainless steel tubular T-joints and 32 stainless steel tubular X-joints was tested with axial compression force applied to the brace members, in which 21 stainless steel tubular X-joints were tested with compressive preload applied to the chord members. All test specimens were fabricated with brace members fully welded at right angle to the center of the continuous chord members.

The welded SHS and RHS consisted of a large range of section sizes. For the chord members, the tubular hollow sections have nominal overall flange width (b_0) ranged from 40 to 200 mm, nominal overall depth of the web (h_0) from 40 to 200 mm, and nominal thickness (t_0) from 1.5 to 6.0 mm. For the brace members, the tubular hollow sections have nominal overall flange width (b_1) ranged from 40 to 150 mm, nominal overall depth of the web (h_1) from 40 to 200 mm, and nominal thickness (t_1) from 1.5 to 6.0 mm. The nominal wall thickness of both chord and brace members go beyond the limits of the current design specifications, in which the nominal wall thickness of hollow sections should not be less than 2.5 mm. The length of the chord member (L_0) was chosen as $h_1 + 5h_0$ to ensure that the stresses at the brace and chord intersection are not affected by the ends of the chord. This is because the points of

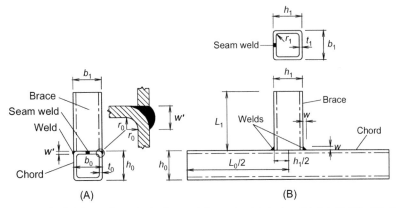

Figure 7.4 Definition of symbols for stainless steel tubular T-joint [7.20]. (A) End view. (B) Elevation.

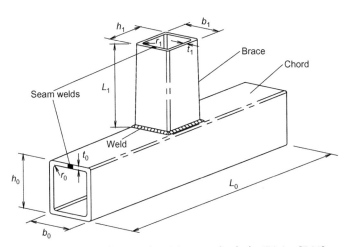

Figure 7.5 Three-dimensional view of stainless steel tubular T-joint [7.20].

contra-flexure on the chord due to the applied load and reactions occur sufficiently far away from the intersection region. The length of the brace member (L_1) was chosen as $2.5h_1$ to avoid the overall buckling of brace members, which cannot reveal the true ultimate capacity of the tubular joints. The measured cross section dimensions of the test specimens are summarized in Refs [7.20,7.21] using the nomenclature defined in Figures 7.4−7.7.

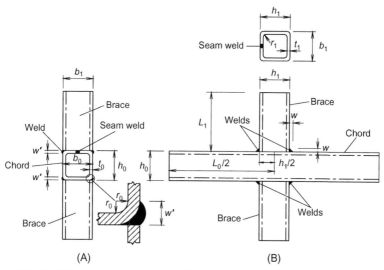

Figure 7.6 Definition of symbols for stainless steel tubular X-joint [7.21]. (A) End view. (B) Elevation.

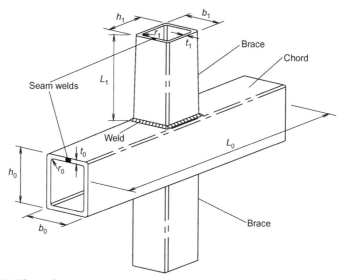

Figure 7.7 Three-dimensional view of stainless steel tubular X-joint [7.21].

7.3.4 Material Properties of Stainless Steel Tubes

As mentioned previously in Section 1.3, the successful test program should carefully measure the material properties of all the test specimen components. Hence, the presented test specimens in this chapter were cold-rolled from austenitic stainless steel type AISI 304 (EN 1.4301), high strength austenitic (HSA), and duplex (EN 1.4462) stainless steel sheets. The stainless steel type AISI 304 is considered as normal strength material, whereas the HSA and duplex are considered as high strength material. The brace and chord members with the same dimensions were selected from the same batch of tubes and so could be expected to have similar material properties. In this study, the stainless steel tubes were obtained from the same batch of specimens conducted by Zhou and Young [7.22] for flexural members. The material properties of the stainless steel tubes were determined by tensile coupon tests, which include the initial Young's modulus (E), the proportional limit stress (σ_p), the static 0.1% ($\sigma_{0.1}$), 0.2% ($\sigma_{0.2}$), 0.5% ($\sigma_{0.5}$), and 1.0% ($\sigma_{1.0}$) tensile proof stresses, the static ultimate tensile stress (σ_u), and the elongation after fracture (ε_f) based on a gauge length of 50 mm.

7.3.5 Test Rig and Procedure
7.3.5.1 Stainless Steel Tubular T-Joints
Once again, as mentioned previously in Section 1.3, the successful test program should carefully look into the details of the test rig, positions, and types of instrumentations as well as the test procedures in order to capture all the significant and required test results. The schematic sketches of the test arrangement presented in this book are shown in Figure 7.8A and B, for the end view and elevation, respectively. Axial compression force was applied to the test specimen by using a servo-controlled hydraulic testing machine. The upper end support was movable to allow tests to be conducted at various specimen dimensions. A special fixed-ended bearing was used at the end of the brace member so that a uniform axial compression load can be applied to the test specimen. The special bearing was connected to the upper end support. The chord member of the test specimen rests on the bottom end plate, which is connected to the bottom support of the testing machine. This provided support to the entire chord member.

Two displacement transducers were positioned on either side of the brace member measuring the vertical deflections at the center of the connecting face of the chord. The transducers were positioned 20 mm away from the faces of the brace member, as shown in Figure 7.9. The flange indentation (u) in the chord member was obtained from the average reading of these two transducers. For the stainless steel tubular T-joint tests, it was observed that the maximum outward deflection (v) of the chord web

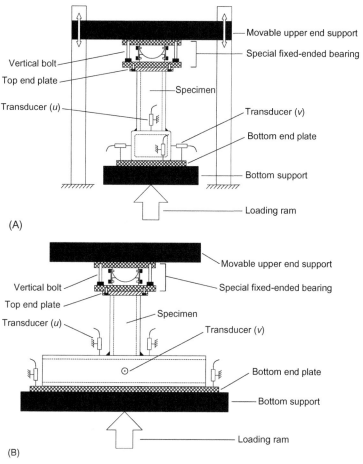

Figure 7.8 Schematic sketch of stainless steel tubular T-joint tests [7.20]. (A) End view. (B) Elevation.

does not occur at the center of the chord sidewall. It may approximately appear near the two-thirds of the overall depth of the chord web (h_0). The exact location of the maximum deformation of the chord sidewall cannot be easily predicted, as it depends on the initial plate imperfection of the chord sidewall. Hence, two displacement transducers were positioned at the center of the chord sidewall to record the deflection. The average of these readings was also taken as the chord web deflection (v), as shown in Figure 7.9. Two other displacement transducers were positioned diagonally on the bottom end plate to measure the axial shortening of the test specimen.

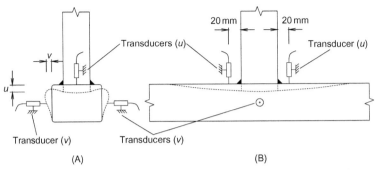

Figure 7.9 Deformations of stainless steel tubular T-joint [7.20]. (A) End view. (B) Elevation.

A 1000-kN capacity servo-controlled hydraulic testing machine was used to apply axial compression force to the test specimen. Displacement control was used to drive the hydraulic actuator at a constant speed of 0.2 mm/min for full width tubular T-joints and 0.4 mm/min for other tubular T-joints. The use of displacement control allowed the tests to be continued in the post-ultimate range. The applied loads and readings of displacement transducers were recorded automatically at regular interval by using a data acquisition system. A photograph of a typical test setup of stainless steel tubular T-joint of specimen TD-C160 × 3-B160 × 3 is shown in Figure 7.10.

7.3.5.2 Stainless Steel Tubular X-Joints Without Chord Preload

Figure 7.11A and B shows the schematic sketches of the test arrangement of stainless steel tubular X-joints without chord preload for the end view and elevation, respectively. The test rig and procedure were employed similarly as previously detailed for the T-joints. The top end plate was bolted to the upper end support, which was considered to be a fixed-ended boundary condition. The load was applied to the bottom brace of the test specimen through a special fixed-ended bearing.

Four displacement transducers were positioned on either side of the brace members measuring the vertical deflections at the center of the connecting faces of the chord. The transducers were positioned 20 mm away from the faces of the brace members, as shown in Figure 7.12. The flange indentation (u) for one face of the chord member was obtained from these transducers. Two displacement transducers were positioned at the center of the chord sidewall to record the deflection. The average of these readings was also taken as the chord web deflection (v), as shown in Figure 7.12.

Figure 7.10 Test setup of stainless steel tubular T-joint of specimen TD-C160 × 3-B160 × 3.

Two other displacement transducers were positioned diagonally on the bottom end plate to measure the axial shortening of the test specimen.

Axial compression force was applied to the braces of the stainless steel tubular X-joints using the same servo-controlled hydraulic testing machine as that used for the tests of stainless steel tubular T-joints. Displacement control was used to drive the hydraulic actuator at a constant speed of 0.2 mm/min for full width tubular X-joints, and 0.4 mm/min for other tubular X-joints. The applied loads and readings of displacement transducers were recorded automatically at regular interval by using the same data acquisition system. A photograph of a typical test setup of stainless steel tubular X-joint without chord preload of specimen XH-C200 × 4-B200 × 4-P0 is shown in Figure 7.13.

7.3.5.3 Stainless Steel Tubular X-Joints with Chord Preload
Similar test rig and nearly the same instrumentation were used for the tests of stainless steel tubular X-joints with chord preload. However, four high strength steel bars were used to apply the compressive preload to the

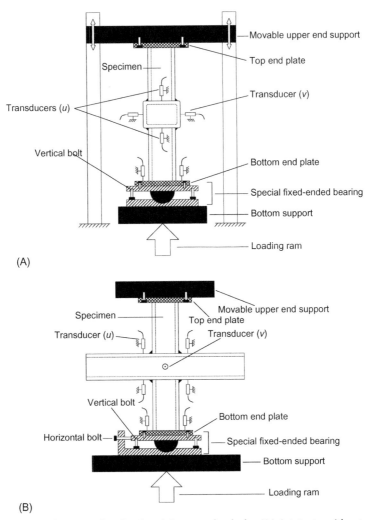

(A)

(B)

Figure 7.11 Schematic sketch of stainless steel tubular X-joint tests without chord preload [7.21]. (A) End view. (B) Elevation.

chord members. The ENERPAC of 1000 kN capacity hydraulic jack was used at one end of the chord members to apply a specified compressive preload (N_p). A load cell was positioned at the other end of the chord member to monitor the applied compressive preload. The schematic sketches of the test arrangement of stainless steel tubular X-joints with chord preload are shown in Figure 7.14A and B for the end view and elevation, respectively. Four single-element strain gauges with a gauge length of 5 mm (TML FLA-5-17) specific for stainless steel were attached at the middle length between the chord end and the face of the brace members

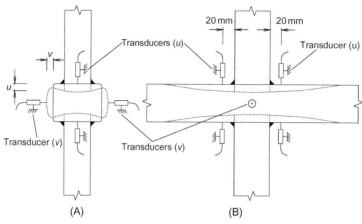

(A)   (B)

Figure 7.12 Deformations of stainless steel tubular X-joint [7.21]. (A) End view. (B) Elevation.

Figure 7.13 Test setup of stainless steel tubular X-joint without chord preload of specimen XH-C200 × 4-B200 × 4-P0.

in order to ensure the preload was uniformly applied to the chord member. The strain gauges were located at the corners of the hollow sections, as shown in Figure 7.14B. The hydraulic jack was continuously adjusted to keep the preload at its initial value ±2% throughout the test.

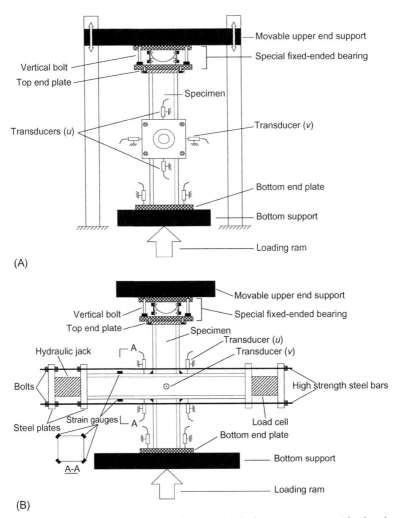

(A)

(B)

Figure 7.14 Schematic sketch of stainless steel tubular X-joint tests with chord pre-load [7.21]. (A) End view. (B) Elevation.

The nominal preloads used in the tests were 10%, 30%, and 50% of the yield load $(A_0\sigma_{0.2})$ of the chord members. Axial compression force was applied to the braces of the test specimen using the same servo–controlled hydraulic testing machine. The same fixed–ended bearing as that used for the tests of stainless steel tubular T-joints was positioned at the upper end to ensure a uniform axial compression force applied to the braces.

Displacement control was used to drive the hydraulic actuator at a constant speed of 0.1 mm/min for full width tubular X-joints, and 0.2 mm/min

Figure 7.15 Test setup of stainless steel tubular X-joint with chord preload of specimen XH-C200 × 4-B200 × 4-P0.1.

for other tubular X-joints. The same measurement system as that used for the tests of stainless steel tubular X-joints without chord preload was employed to obtain the chord flange indentation (u), the chord web deflection (v), and the axial shortening of the test specimens. The same data acquisition system was also used during the tests to record the applied loads and readings of displacement transducers manually when the compressive preloads were within the range of the specified values. A photograph of a typical test setup of stainless steel tubular X-joint with chord preload of specimen XH-C200 × 4-B200 × 4-P0.1 is shown in Figure 7.15.

7.4. FINITE ELEMENT MODELING OF METAL TUBULAR CONNECTIONS

7.4.1 General

The test results given in Refs [7.20,7.21] have provided enough information for the verification of finite element models developed by Feng and Young [7.23]. The developed models for cold-formed stainless steel tubular

T- and X-joints of SHS and RHS used the general-purpose finite element program ABAQUS [1.27]. Three finite element models were developed, namely, the T-joints, X-joints, and X-joints with chord preload, including various critical influential factors, such as modeling of materials and welds, contact interaction between the T-joint specimens and the supporting plate, as well as loading and boundary conditions. The load—displacement nonlinear analysis was performed by using the (*STATIC) procedure available in the ABAQUS [1.27] library. Both geometric and material nonlinearities have been taken into account in the finite element models. The element type and mesh size of the stainless steel tube and the welding material were carefully determined by the convergence studies to provide accurate results with reasonable computational cost. The joint strengths, failure modes, and load—deformation curves of stainless steel tubular joints were all obtained from the finite element analysis.

A 4-node doubly curved shell element with reduced integration (S4R) has been used by many researchers to model the brace and chord members of welded tubular joints. A 5-point integration was applied through the shell thickness with full complement of six degrees of freedom per node. The S4R element provides accurate solution to most applications by allowing for transverse shear deformation, which is important for the simulation of thick shell element. However, the shell element in the contact algorithm could incorrectly allow penetration of one member into the other due to its ignorance of physical thickness of the element. Therefore, three-dimensional 8-node solid element with reduced integration (C3D8R) was used in this study to model the cold-formed stainless steel tubular joints. This element is fully isoparametric with first-order interpolation. The use of solid element rather than shell element for the modeling of welded tubular joints could achieve accurate results with slight increase of computational time.

In order to obtain the optimum finite element mesh size, the convergence studies were carried out. It was found that the mesh size of approximately 3×3 mm (length by width) for small specimens, 6×6 mm for medium specimens, and 10×10 mm for large specimens in modeling the flat portions of both flange and web elements could achieve accurate results with the minimum computational time. The corresponding length-to-width ratio of the elements is equal to 1.0 for both brace and chord members. A finer mesh of four elements was used at the corner portions due to their importance in transferring the stress from the flange to the web. Based on the study of Choo et al. [7.12], four layers of solid elements were employed across the tube wall thickness for thick-walled tubular members ($b_0/t_0 \leq 20$ for chord member; and $b_1/t_1 \leq 20$ for brace member), while two layers of solid elements were used for thin-walled tubular members ($b_0/t_0 > 20$ for chord member; and $b_1/t_1 > 20$ for brace member). This technique was implemented through the thickness of all tubular members to provide satisfactory simulation of possible nonlinearity in the thickness direction in this study.

The measured stress—strain curves of stainless steel tubes were incorporated in the finite element models. The static stress—strain curves were determined using the static loads near the proportional limit stress and the ultimate tensile stress. The initial part of the nonlinear stress—strain curve represents the elastic property up to the proportional limit stress with measured Young's modulus (E), and Poisson's ratio (ν) equals to 0.3. The nonlinear finite element analysis involved large plastic strains; therefore, the static stress—strain curves obtained from the tensile coupon tests were converted to the true stress—logarithmic plastic strain curves. The material nonlinearity behavior was included in the finite element models. The true stress (σ_{true}) and logarithmic plastic strain (ε_{ln}^{pl}) were calculated based on the recommendation of ABAQUS standard user's manual [1.27].

The material properties of the welds were adopted from the material properties of electrodes as described in the AWS A5.11 specification [7.24]. Some of the researchers omitted the welds for simplicity in the modeling of welded tubular joints, which led to underestimation of joint strengths, whereas Van der Vegte et al. [7.25] modeled the weld geometry using a "ring" of shell element along the brace and chord intersection region. However, an unrealistic air-filled void was introduced into the finite element models [7.26]. In this study, three-dimensional 8-node solid element with reduced integration (C3D8R) was employed to simulate the welds at brace and chord intersection region. The weld geometry was modeled in accordance with the measured dimensions of welds for stainless steel tubular T-joints, X-joints, and X-joints with chord preload.

7.4.2 Stainless Steel Tubular T-Joints

For the modeling of stainless steel tubular T-joints, the top end of brace member and the loaded end at the bottom supporting plate were restrained against all degrees of freedom, except for the displacement at the loaded end in the direction of the applied load. The nodes other than the loaded end and the top end were free to translate and rotate in any directions. The load was applied in increments by using the (*STATIC) method available in the ABAQUS [1.27] library. The static uniform loads were applied by

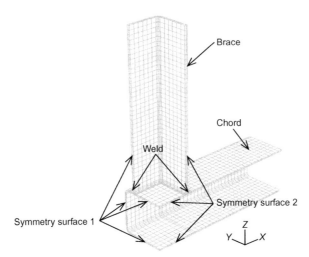

Figure 7.16 Finite element mesh of stainless steel tubular T-joint [7.23].

means of displacement at each node of the loaded end, which was identical to the experimental test setup. The nonlinear geometry parameter (*NLGEOM) was used for the consideration of large displacement analysis. The contact between the bottom supporting plate and the bottom surface of chord member plays an important role in the load transferring mechanism. Therefore, an analytic rigid-deformable contact interaction between the rigid bottom supporting plate and the deformable chord member was established using a "master–slave" algorithm available in the ABAQUS [1.27] library. The contact interaction allows the surfaces to separate under the effect of tensile force. However, they are not allowed to penetrate each other under the effect of compressive force. In the nonlinear analysis, an initial load step was applied to establish the initial contact interaction before the application of loading. A quarter of tubular joint was modeled by taking advantage of two planes of symmetry in geometry, loading application, and boundary conditions. The nodes on symmetry surfaces 1 and 2 were restrained against displacements in X and Y directions, respectively, due to symmetry, as shown in Figure 7.16 for stainless steel tubular T-joint.

7.4.3 Stainless Steel Tubular X-Joints without Chord Preload

In the modeling of stainless steel tubular X-joints without chord preload, the top and bottom ends of brace members were restrained against all degrees of freedom, except for the displacement at the loaded end in the direction of the applied load. The nodes other than the top and bottom ends were free to translate and rotate in any directions. A similar

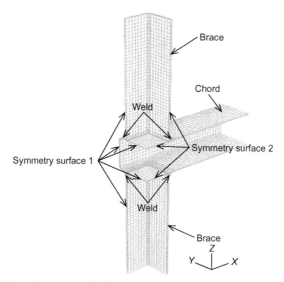

Figure 7.17 Finite element mesh of stainless steel tubular X-joint [7.23].

procedure as the tubular T-joints was used to apply the static uniform loads. A quarter of tubular joint was also modeled. Figure 7.17 shows the finite element mesh of stainless steel tubular X-joint.

7.4.4 Stainless Steel Tubular X-Joints with Chord Preload

As for the modeling of stainless steel tubular X-joints with chord preload, two consecutive steps were required to complete the nonlinear finite element analysis. In the first step, the chord member was directly preloaded using the (*STATIC) method to the prescribed load level. This preload level was maintained constant throughout the full analysis. In the second step, the brace member was loaded by means of displacement using the (*STATIC) method. The loading method applied to the chord and brace members was identical to the test procedure.

7.5. VERIFICATION OF FINITE ELEMENT MODELS

A comparison between the experimental and numerical results was carried out to verify the finite element models. A total of 20 T-joints, 10 X-joints, and 19 X-joints with chord preload for cold-formed stainless steel SHS and RHS tubes was analyzed. The joint strengths, load—axial shortening curves, and deformed shapes based on the different failure modes of stainless steel tubular T- and X-joints have been investigated. The joint strengths obtained from the tests (P_{Test}) were compared with

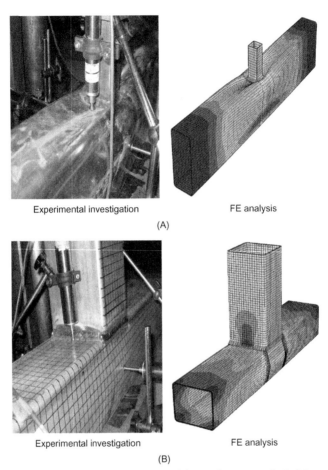

Experimental investigation   FE analysis

(A)

Experimental investigation   FE analysis

(B)

Figure 7.18 Comparison of experimental and finite element analysis failure modes for stainless steel tubular T-joints [7.23]. (A) Chord face failure. (B) Chord sidewall failure.

the finite element analysis results (P_{FE}). The mean values of the P_{Test}/P_{FE} ratio are 0.99, 0.99, and 1.00 with the corresponding coefficients of variation of 0.066, 0.062, and 0.068 for stainless steel tubular T-joints, X-joints, and X-joints with chord preload, respectively. Good agreement between the experimental and finite element analysis results was achieved.

Three different failure modes including chord face failure, chord sidewall failure, and local buckling failure of brace, which were observed in the experimental investigation, were also verified by the finite element models. The chord face failure and chord sidewall failure observed in the tests were closely simulated by the finite element analysis, as shown in Figures 7.18 and 7.19 for stainless steel tubular T- and X-joints,

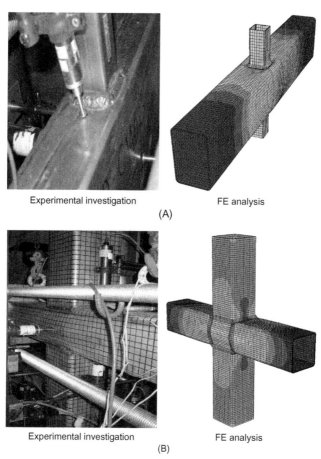

Experimental investigation   FE analysis
(A)

Experimental investigation   FE analysis
(B)

Figure 7.19 Comparison of experimental and finite element analysis failure modes for stainless steel tubular X-joints [7.23]. (A) Chord face failure. (B) Chord sidewall failure.

respectively. However, the failure mode of local buckling failure of brace is similar to that observed in stub column test, which cannot reveal the true ultimate capacity of welded tubular joints. Thus, this failure mode was not investigated in this study.

The load—axial shortening curves obtained from experimental and finite element analysis were compared in Figures 7.20—7.22 for stainless steel tubular T-joints, X-joints, and X-joints with chord preload, respectively. Both high strength stainless steel (duplex and high strength austenitic) and normal strength stainless steel (AISI 304) tubular joints were

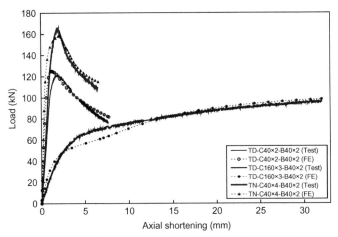

Figure 7.20 Comparison of experimental and finite element analysis load—axial shortening curves for stainless steel tubular T-joints [7.23].

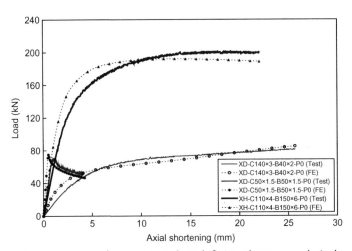

Figure 7.21 Comparison of experimental and finite element analysis load—axial shortening curves for stainless steel tubular X-joints without chord preload [7.23].

included in the comparison. Good agreement between the experimental and finite element analysis results was obtained for all typical failure modes. It was demonstrated that the newly developed finite element models successfully predicted the structural performance of cold-formed stainless steel tubular T-joints, X-joints, and X-joints with chord preload.

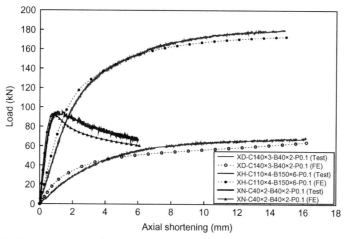

Figure 7.22 Comparison of experimental and finite element analysis load–axial shortening curves for stainless steel tubular X-joints with chord preload [7.23].

7.6. SUMMARY

This chapter started by presenting the experimental investigation of cold-formed stainless steel tubular T- and X-joints of SHS and RHS, which were chosen as examples in this book. The test specimens cold-rolled from high strength and normal strength stainless steel materials consisted of different brace width to chord width ratio (β) ranging from 0.5 to 1.0, brace thickness to chord thickness ratio (τ) from 0.5 to 2.0, chord width to chord thickness ratio (2γ) from 10 to 50, and different levels of compressive preload (10%, 30%, and 50% of the yield load of the chord members) applied to the chord members of stainless steel tubular X-joints. A special fixed-ended bearing was used to ensure that the pure axial compression force without any bending moment was applied to the test specimens. The test procedure and data acquisition system were carefully designed to record the test data. The joint strengths, failure modes involving chord face failure, chord sidewall failure, and local buckling failure of brace, as well as the load–deformation curves for all test specimens were obtained. The tests on cold-formed stainless steel tubular T- and X-joints are detailed in Refs [7.20,7.21].

This chapter also presented the finite element analysis of cold-formed stainless steel tubular T-joints, X-joints, and X-joints with chord preload. Both high strength stainless steel (duplex and high strength austenitic) and normal strength stainless steel (AISI 304) tubular joints were simulated.

The geometric and material nonlinearities of stainless steel tubular joints were included in the finite element models. The element type and mesh size of the stainless steel tube and welding material were carefully determined by the convergence studies to provide accurate results with reasonable computational cost. Various critical influential factors, such as modeling of materials, welds, contact interaction, and loading and boundary conditions were all taken into consideration in the finite element analysis. The joint strengths, failure modes, and load−deformation curves were all obtained from the finite element analysis.

REFERENCES

[7.1] I-DEAS. TMG Reference Manual. USA: MAYA Heat Transfer Technologies Ltd., 2003.
[7.2] MARC. User's guide. California, USA: MARC Analysis and Research Corporation, 2005.
[7.3] PAFEC. Data preparation. Nottingham, UK: Publication of PAFEC Limited, Strelley Hall, 1989.
[7.4] MSC/PATRAN version 2000 r2. User's manual. California, USA: The MacNeal-Schwendler Corporation, 2000.
[7.5] Packer, J. A. Parametric study of rectangular hollow section welded joints. Proceedings of the 8th Australasian Conference on the Mechanics of Structures and Materials, Newcastle, Australia, 29.1−29.6, 1982.
[7.6] Packer, J. A., Korol, R. M., Mirza, F. A., and Ostrowski, P.K. Finite element study of the flexibility of rectangular hollow section truss gapped K-joints. Proceedings of the 9th Australasian Conference on the Mechanics of Structures and Materials, Sydney, Australia, 268−272, 1984.
[7.7] Ebecken, Nelson. F. F., De Lima, Edison. C. P., Landau, Luiz., Chueiri, Lauro. H. M. and Benjamin, Adilson. C. Computational techniques for the elastic-plastic analysis of tubular joints. Engineering Computations, 4(2), 126−130, 1987.
[7.8] Zhang, Z. L., Shen, Z. Y., and Chen, X. C. Nonlinear FEM analysis and experimental study of ultimate capacity of welded RHS joints. Proceedings of the 3rd International Symposium on Tubular Structures, Lappeenranta, Finland, 232−240, 1989.
[7.9] Moffat, D. G., Kruzelecki, J. and Blachut, J. The effects of chord length and boundary conditions on the static strength of a tubular T-joint under brace compression loading. Marine Structures, 9(10), 935−947, 1996.
[7.10] Saidani, M. The effect of joint eccentricity on the distribution of forces in RHS lattice girders. Journal of Constructional Steel Research, 47(3), 211−221, 1998.
[7.11] Lee, M. M. K. Strength, stress and fracture analyses of offshore tubular joints using finite elements. Journal of Constructional Steel Research, 51(3), 265−286, 1999.
[7.12] Choo, Y. S., Qian, X. D., Liew, J. Y. R. and Wardenier, J. Static strength of thick-walled CHS X-joints—Part I. New approach in strength definition. Journal of Constructional Steel Research, 59(10), 1201−1228, 2003.
[7.13] Karamanos, S. A. and Anagnostou, G. Pressure effects on the static response of off-shore tubular connections. Marine Structures, 17(6), 455−474, 2004.
[7.14] Gho, W. M., Yang, Y. and Gao, F. Failure mechanisms of tubular CHS joints with complete overlap of braces. Thin-Walled Structures, 44(6), 655−666, 2006.

[7.15] Van Der Vegte, G. J., Makino, Y., and Wardenier, J. The effect of chord load on the ultimate strength of CHS X-joints. Proceedings of the 17th International Offshore and Polar Engineering Conference, Lisbon, Portugal, 3387–3394, 2007.

[7.16] Gho, W. M. and Yang, Y. Parametric equation for static strength of tubular circular hollow section joints with complete overlap of braces. *Journal of Structural Engineering, ASCE,* 134(3), 393–401, 2008.

[7.17] Shao, Y. B., Lie, S. T. and Chiew, S. P. Static strength of tubular T-joints with reinforced chord under axial compression. Advances in Structural Engineering, 13 (2), 369–377, 2010.

[7.18] Van der Vegte, G. J., Wardenier, J. and Puthli, R. S. FE analysis for welded hollow-section joints and bolted joints. Proceedings of the Institution of Civil Engineers: Structures and Buildings, 163(6), 427–437, 2010.

[7.19] Liu, Y. and Deng, P. FEM analysis of CHS X-joints on the effect of out-of-plane bending performance. International Conference on Multimedia Technology, ICMT, 3968–3971, 2011.

[7.20] Feng, R. and Young, B. Experimental investigation of cold-formed stainless steel tubular T-joints. Thin-Walled Structures, 46(10), 1129–1142, 2008.

[7.21] Feng, R. and Young, B. Tests and behaviour of cold-formed stainless steel tubular X-joints. Thin-Walled Structures, 48(12), 921–934, 2010.

[7.22] Zhou, F. and Young, B. Tests of cold-formed stainless steel tubular flexural members. Thin-Walled Structures, 43(9), 1325–1337, 2005.

[7.23] Feng, R. and Young, B. Design of cold-formed stainless steel tubular T- and X-joints. Journal of Constructional Steel Research, 67(3), 421–436, 2011.

[7.24] American Welding Society. Specification for nickel and nickel-alloy welding electrodes for shielded metal arc welding. Miami, Florida, USA: AWS A5.11/A5.11M, 2005.

[7.25] Van der Vegte, G. J., de Koning, C. H. M., Puthli, R. S. and Wardenier, J. Numerical simulation of experiments on multiplanar tubular steel X-joints. International Journal of Offshore and Polar Engineering, 1(3), 200–207, 1991.

[7.26] Van der Vegte, G. J., Choo, Y. S., Liang, J. X., Zettlemoyer, N. and Liew, J. Y. R. Static strength of T-joints reinforced with doubler or collar plates. II: Numerical simulations. *Journal of Structural Engineering, ASCE,* 131(1), 129–138, 2005.

CHAPTER 8

Design Examples of Metal Tubular Connections

8.1. GENERAL REMARKS

Chapter 7 has presented different finite element models for the innovative metal tubular connections presented as examples. The presented models were verified against the corresponding experimental investigations on cold-formed stainless steel tubular T- and X-joints. The chapter has detailed the developing of the models based on the background previously detailed in Chapters 1–4. This chapter highlights the design rules specified in current codes of practice for the connections as well as the design examples of the connections. The main objective is to assess the findings of the finite element models against current codes of practice and to provide a complete piece of work on detailing the design procedures for metal tubular connections. The aforementioned objective is the main goal of this book. This chapter details the parametric studies performed using the verified finite element models. The parametric studies have accounted for the effects of the change in main parameters affecting the strength and behavior of cold-formed stainless steel tubular joints. The joint strengths, failure modes, and load–deformation curves of cold-formed stainless steel tubular T- and X-joints obtained experimentally and numerically were highlighted and discussed to show the effectiveness of the results. The design rules specified for this form of construction were also presented in the chapter. In addition, this chapter presents the comparison between the experimental/finite element analysis results and the predictions obtained from current design specifications for carbon steel and stainless steel structures, as well as the proposed design rules for cold-formed stainless steel tubular joints. Furthermore, some design examples are presented in this chapter to demonstrate the design procedures of cold-formed stainless steel tubular T-, X-, and X-joint with chord preload' with chord preload subjected to different failure modes using the current and proposed design rules.

Overall, this chapter shows the effectiveness and usefulness of the finite element modeling of tubular joints in improving design rules specified in current codes of practice and providing better understanding of the structural behavior and performance of this form of structures, which is one of the main objectives of this book.

8.2. PARAMETRIC STUDY OF METAL TUBULAR CONNECTIONS

8.2.1 General

The number of tests carried out in any test program depends on many factors, including cost, labor, time, and facilities. Therefore, the number of tests conducted on any form of construction, indeed, remains limited. The effectiveness of the finite element models is that, once validated and verified, they can be used to perform numerous analyses and parametric studies highlighting the effects of the change in different parameters affecting the structural performance of the form of construction. For the joints highlighted in this chapter, numerical analyses were performed by Feng and Young [7.23] for cold-formed stainless steel square and rectangular hollow sections (SHS and RHS) tubular T-, X-, and X-joints with chord preload by using the general-purpose finite element program ABAQUS [1.27], which was described in detail in Chapter 7. The nonlinear finite element models were calibrated against the corresponding experimental investigation as introduced in Chapter 7. An extensive parametric study was carried out using the verified finite element models to investigate the strength and behavior of cold-formed stainless steel tubular T- and X-joints. The joint strengths, failure modes, and load—deformation curves of stainless steel tubular joints were all obtained from the numerical investigation.

8.2.2 Specimens Investigated in the Parametric Study

The parametric study, highlighted in this chapter, was carried out using the verified finite element models, detailed in Chapter 7, to evaluate the effects of geometric parameters and compressive chord preload on the structural behavior of cold-formed stainless steel tubular joints. A total of 122 T-joints, 20 X-joints, and 30 X-joints with chord preload of cold-formed stainless steel SHS and RHS tubes was analyzed in the parametric study. The SHS and RHS consisted of a large range of section sizes, which were selected from the range of practical applications. For the chord members, the tubular hollow sections have overall flange width (b_0) ranged from 25 to 300 mm, overall depth of the web (h_0) from 30 to 400 mm, and thickness (t_0) from 1 to 16 mm. For the brace members, the tubular hollow sections have overall flange width (b_1) ranged from 25 to 300 mm, overall depth of the web (h_1) from 25 to 400 mm, and thickness (t_1) from 1 to 16 mm. In this study, the wall thicknesses of both chord and brace members go beyond the limits of the current design specifications, in which the wall thickness of hollow sections should not be less than 2.5 mm. The length of the chord member (L_0) was chosen as $h_1 + 5h_0$ to ensure that the stresses at the brace and chord intersection are not affected by the ends of the chord member. The length of the brace member (L_1) was chosen as $2.5h_1$ to avoid the overall buckling of brace member. The external corner radius (R_i) of stainless steel tube was taken as $2.5t$, provided the thickness of tube (t) is larger than 3 mm; otherwise, the external corner radius was taken as $2t$, which was recommended in the AISC design guideline [8.1]. The weld size (w) was taken as $2t$ based on the recommendation in the American Welding Society (AWS) D1.1/D1.1M specification [8.2], where t is the thickness of thinner part between brace and chord members. The compressive chord preloads applied to the tubular X-joints were taken as 10%, 30%, and 50% of the yield load ($A_0\sigma_{0.2}$) of chord members, which was identical to the experimental investigation.

8.2.3 Influential Factors

The effects of critical influential factors on the strength and behavior of cold-formed stainless steel tubular joints were evaluated, which include the brace width to chord width ratio ($\beta = b_1/b_0$), brace thickness to chord thickness ratio ($\tau = t_1/t_0$), chord width to chord thickness ratio ($2\gamma = b_0/t_0$), brace width to brace thickness ratio (b_1/t_1), chord depth to chord thickness ratio (h_0/t_0), brace depth to brace thickness ratio (h_1/t_1), chord depth to chord width ratio (h_0/b_0), brace depth to brace width ratio (h_1/b_1), and the compressive chord preload (N_p). The validity range of these geometric parameters defined in the tests and parametric study were purposely designed beyond those given in the current design specifications for carbon steel and stainless steel structures. The joint strengths, failure modes involved chord face failure, chord sidewall failure, and local buckling failure of brace, as well as the load−deformation curves of stainless steel tubular joints were all summarized in Ref. [7.23].

8.3. DESIGN RULES OF METAL TUBULAR CONNECTIONS

8.3.1 Current Design Specifications

A complete piece of work concerning finite element modeling of a metal structure should assess the numerical findings against current codes of practice. This is to provide useful guides and recommendations for researchers, designers, and practitioners interested in the behavior and design of the metal structure. For the tubular joints highlighted in this chapter, the design rules for carbon steel SHS and RHS tubular joints are available in the Comité International pour le Développement et l'Étude de la Construction Tubulaire (CIDECT) Design Guide No. 3 [8.3]. The Eurocode 3 part 1.8 [8.4] provides similar design guideline for carbon steel SHS and RHS tubular joints by adopting the CIDECT recommendations. The design guidelines presented in these publications have been incorporated in the updated International Institute of Welding (IIW) document [8.5]. The Australian/ New Zealand Standard (AS/NZS 4673) [8.6] is the only design specification currently used for cold-formed stainless steel tubular joints. The design rules given in this specification are generally adopted from the CIDECT recommendations for carbon steel tubular joints. Since there is little research being carried out in this area, some design criteria for stainless steel tubular joints are not covered in this design specification. The Australian/ New Zealand Standard (AS/NZS 4673) [8.6] does not have design equation for stainless steel RHS tubular joints. The design strengths (N_1) of stainless steel SHS and RHS tubular joints can be calculated using the following design equations based on different failure modes:

For $\beta \leq 0.85$ (chord face failure):

$$N_1 = \frac{f_{y0}t_0^2}{(1 - \beta)\sin \theta_1} \left[\frac{2\eta}{\sin \theta_1} + 4(1 - \beta)^{0.5} \right] f(n) \quad \text{(CIDECT No. 3[8.3])}$$

(8.1)

$$f(n) = 1.3 + \frac{0.4n}{\beta} \leq 1.0$$

(8.2)

$$n = \frac{N_\text{p}}{f_{y0}A_0} < 0$$

(8.3)

$$N_1 = \frac{k_n f_{y0}t_0^2}{(1 - \beta)\sin \theta_1} \left(\frac{2\eta}{\sin \theta_1} + 4\sqrt{1 - \beta} \right) / \gamma_{\text{M5}} \quad \text{(EC3 part 1.8 [8.4])}$$

(8.4)

$$\varphi N_{1n} = \frac{f_{y0} t_0^2}{(1 - \beta)\sin \theta_1} \left[\frac{2\beta}{\sin \theta_1} + 4(1 - \beta)^{0.5} \right] k_n \left(\frac{\varphi}{0.9} \right)$$

(8.5)

(AS/NZS 4673 [8.6] for SHS)

$$k_n = 1.3 - \frac{0.4n}{\beta} \leq 1.0$$

(8.6)

$$n = \frac{N_p}{f_{y0} A_0} > 0$$

(8.7)

For $\beta = 1.0$ (chord sidewall failure):

$$N_1 = \frac{f_k t_0}{\sin \theta_1} \left[\frac{2h_1}{\sin \theta_1} + 10t_0 \right] \quad \text{(CIDECT No. 3 [8.3])}$$

(8.8)

$$N_1 = \frac{f_b t_0}{\sin \theta_1} \left(\frac{2h_1}{\sin \theta_1} + 10t_0 \right) / \gamma_{M5} \quad \text{(EC3 part 1.8 [8.4])}$$

(8.9)

For $\beta \geq 0.85$ (local buckling failure of brace):

$$N_1 = f_{y1} t_1 [2h_1 - 4t_1 + 2b_e] \quad \text{(CIDECT No. 3 [8.3])}$$

(8.10)

$$N_1 = f_{y1} t_1 (2h_1 - 4t_1 + 2b_{eff}) / \gamma_{M5} \quad \text{(EC3 part 1.8 [8.4])}$$

(8.11)

where f_{y0} is the yield stress of the chord, f_{y1} is the yield stress of the brace, t_0 is the thickness of the chord, t_1 is the thickness of the brace, β is the ratio of brace width to chord width (b_1/b_0), η is the ratio of brace depth to chord width (h_1/b_0), θ_1 is the angle between the brace and the chord, A_0 is the cross-sectional area of the chord, ϕ is the resistance factor, γ_{M5} is the partial safety factor, N_p is the compressive preload applied to the chord, n is the preload ratio in the chord, k_n and $f(n)$ are the parameters that account for the influence of compression chord longitudinal stresses, f_k and f_b are the chord sidewall flexural buckling stresses, b_e and b_{eff} are the effective widths of the brace.

For $0.85 \leq \beta \leq 1.0$:

According to the CIDECT Design Guide No. 3 [8.3], the linear interpolation between the governing value for chord face failure at $\beta = 0.85$ and the governing value for chord sidewall failure at $\beta = 1.0$ should be used.

The design strengths given in the CIDECT Design Guide No. 3 [8.3] have already incorporated the resistance factor (ϕ), which are the products of the nominal strengths and the resistance factor. The design strengths given in the Eurocode 3 part 1.8 [8.4] include the partial safety factor (γ_{M5}), where $\gamma_{M5} = 1.0$ is recommended. This partial safety factor should be shown in the design equations, although it seems that it has no influence on the design strengths. However, the value may be modified in revising the design rules in the future. The design strengths given in the Australian/New Zealand Standard (AS/NZS 4673) [8.6] for cold-formed stainless steel structures include the strength reduction factor (ϕ), where $\phi = 0.9$ is adopted. By comparing the design strength equations for welded tubular joints given in these design specifications, it can be generally concluded that the same design strengths can be obtained. According to the recommendation proposed by Rasmussen and Young [8.7], the nominal strength (N_{1n}) can be calculated as $N_{1n} = 1.1 \times N_1$, where N_1 is the design strength given in the CIDECT Design Guide No. 3 [8.3]. For the full-width tubular joints in compression ($\beta = 1$), the design provisions require calculation of the chord sidewall flexural buckling stress. The buckling curve "a" having the imperfection factor $\alpha = 0.21$ in the Eurocode 3 part 1.1 [8.8] has been chosen for calculating the chord sidewall flexural buckling stress.

8.3.2 Proposed Design Rules

In order to improve the design rules specified in current codes of practice, a design formula was proposed by Feng and Young [7.23] for cold-formed stainless steel SHS and RHS tubular T- and X-joints, which were fabricated from austenitic stainless steel type AISI 304, high-strength austenitic (HSA), and duplex (EN 1.4462) stainless steel. The reduction factors specific to different failure modes were introduced based on the current design specifications. The validity range of some critical geometric parameters in the proposed design formulae are beyond those defined in the current design specifications. The proposed nominal strengths (N_{1np}) of stainless steel SHS and RHS tubular joints can be calculated using the following design equations based on different failure modes:

For $0.2 \leq \beta < 0.7$ (chord face failure):

$$N_{1np} = \alpha_A N_{1n} = \alpha_A N_1 \times 1.1 \quad \text{(Feng and Young [7.23])} \quad (8.12)$$

$$N_1 = \frac{f_{y0}t_0^2}{(1-\beta)\sin\theta_1}\left[\frac{2\eta}{\sin\theta_1} + 4(1-\beta)^{0.5}\right]f(n) \quad \text{(CIDECT No. 3 [8.3])}$$

$$(8.13)$$

$$\alpha_A = 1 - \frac{b_0}{100 t_0} \tag{8.14}$$

$$f(n) = 1 - \frac{0.1n}{\beta} \tag{8.15}$$

$$n = \frac{N_P}{f_{y0} A_0} > 0 \tag{8.16}$$

For $\beta > 0.85$ (chord sidewall failure):

$$N_{1np} = \alpha_B N_{1n} = \alpha_B N_1 \times 1.1 \quad \text{(Feng and Young [7.23])} \tag{8.17}$$

$$N_1 = \frac{f_k t_0}{\sin \theta_1} \left[\frac{2 h_1}{\sin \theta_1} + 10 t_0 \right] \quad \text{(CIDECT No. 3 [8.3])} \tag{8.18}$$

$$\alpha_B = \frac{2}{25} \left(\frac{h_0}{t_0} - 1 \right) \tag{8.19}$$

For $0.7 \leq \beta \leq 0.85$ (combined chord face failure and chord sidewall failure):

$$N_{1np} = \alpha_{A+B} N_{1n} = \alpha_{A+B} N_1 \times 1.1 \quad \text{(Feng and Young [7.23])} \tag{8.20}$$

$$N_1 = \frac{f_{y0} t_0^2}{(1 - \beta) \sin \theta_1} \left[\frac{2\eta}{\sin \theta_1} + 4(1 - \beta)^{0.5} \right] f(n) \quad \text{(CIDECT No. 3 [8.3])} \tag{8.21}$$

$$\alpha_{A+B} = 1 + \frac{3 b_0}{1000 t_0} \tag{8.22}$$

$$f(n) = 1 - \frac{0.1n}{\beta} \tag{8.23}$$

$$n = \frac{N_P}{f_{y0} A_0} > 0 \tag{8.24}$$

For $\beta \geq 0.85$, $\tau \leq 0.55$, and $b_e \geq b_1$ (local buckling failure of brace):

$$N_{1n} = 1.1 \times N_1 \quad \text{(Feng and Young [7.23])} \tag{8.25}$$

$$N_1 = f_{y1} t_1 [2 h_1 - 4 t_1 + 2 b_e] \quad \text{(CIDECT No. 3 [8.3])} \tag{8.26}$$

8.4. COMPARISON OF EXPERIMENTAL AND NUMERICAL RESULTS WITH DESIGN CALCULATIONS

It should be noted that the design equations (8.1−8.3), (8.8), (8.10), (8.13), (8.16), (8.18), (8.21), (8.24), and (8.26) are given in the CIDECT Design Guide No. 3 [8.3], the design equations (8.4), (8.6), (8.7), (8.9), and (8.11) are given in the Eurocode 3 part 1.8 [8.4], the design equation (8.5) is given in the Australian/New Zealand Standard (AS/NZS 4673) [8.6], and the design equations (8.12), (8.14), (8.15), (8.17), (8.19), (8.20), (8.22), (8.23), (8.25) were proposed by Feng and Young [7.23]. The failure loads (N_f) obtained from the experimental and numerical investigations were compared with the nominal strengths (N_{1n}) calculated using the design formulae given in the CIDECT Design Guide No. 3 [8.3]. The mean values of failure load to nominal strength ratio N_f/N_{1n} are 0.68, 2.76, 1.11, and 1.02, with the corresponding coefficient of variation (COV) of 0.354, 0.628, 0.217, and 0.098 for stainless steel tubular T- and X-joints subjected to chord face failure, chord sidewall failure, combined chord face failure and chord sidewall failure, and local buckling failure of brace, respectively. In addition, the failure loads (N_f) obtained from the experimental and numerical investigations were also compared with the proposed nominal strengths (N_{1np}) calculated using the design formulae proposed by Feng and Young [7.23]. A good agreement was obtained with the mean values of failure load to proposed nominal strength ratio N_f/N_{1np} of 1.01, 1.01, and 1.01, and the corresponding COV of 0.269, 0.282, and 0.217 for stainless steel tubular T- and X-joints subjected to chord face failure, chord sidewall failure, and combined chord face failure and chord sidewall failure, respectively.

It can be generally concluded from the comparison that the design rules specified in the current design specifications are quite unconservative for stainless steel tubular T- and X-joints subjected to chord face failure, whereas it is quite conservative for specimens failed by chord sidewall. The comparison also shows that the design rules specified in the current design specifications are slightly conservative for stainless steel tubular T- and X-joints subjected to combined chord face failure and chord sidewall failure, whereas it is generally appropriate for specimens failed by local buckling of brace. It was shown that the proposed design formulae are generally much more accurate than those given in the current design specifications.

8.5. DESIGN EXAMPLES

8.5.1 General

Some design examples are presented to illustrate the design procedures of cold-formed stainless steel tubular T-, X-, and X-joints with chord pre-load subjected to different failure modes, which are selected from the parametric study as described in detail in Section 8.2. The joint strengths calculated using the design equations given in the CIDECT Design Guide No. 3 [8.3] and proposed by Feng and Young [7.23] are compared with those obtained from the numerical investigation of cold-formed stainless steel tubular T- and X-joints by using the verified finite element models developed in Chapter 7, which were summarized in [7.23].

8.5.2 Stainless Steel Tubular T-Joint

●●●
_____

Design Example 1

Determine the joint strength of stainless steel tubular T-joint of TD-C160 × 80 × 3-B40 × 40 × 2. Given that

RHS chord: 160 × 80 × 3 ($h_0 = 160.5$ mm, $b_0 = 80.6$ mm, $t_0 = 2.96$ mm, $r_0 = 6.00$ mm, $R_0 = 8.96$ mm, $A_0 = 1354$ mm^2)

SHS brace: 40 × 40 × 2 ($h_1 = 40.1$ mm, $b_1 = 40.3$ mm, $t_1 = 1.96$ mm, $r_1 = 2.00$ mm, $R_1 = 3.96$ mm, $A_1 = 290$ mm^2)

$E_0 = 208$ GPa, $E_1 = 216$ GPa, $f_{y0} = 536$ MPa, $f_{y1} = 707$ MPa, $\theta_1 = 90°$, and $\sin \theta_1 = 1.0$.

Solution 1: (CIDECT No. 3 [8.3])

$$\beta = b_1/b_0 = 40.3/80.6 = 0.50 < 0.85 \rightarrow \text{Chord face failure}$$

Equation (8.1) given in the CIDECT Design Guide No. 3 [8.3] is used as follows:

$$N_1 = \frac{f_{y0} t_0^2}{(1 - \beta)\sin \theta_1} \left[\frac{2\eta}{\sin \theta_1} + 4(1 - \beta)^{0.5} \right] f(n)$$

where

$$\eta = h_1/b_0 = 40.1/80.6 = 0.50, \quad f(n) = 1.3 - \frac{0.4n}{\beta} \le 1.0$$

in which

$$n = \frac{N_p}{A_0 f_{y0}} = 0, \quad \text{thus } f(n) = 1.0$$

Therefore,

$$N_1 = 35.9 \text{ kN}, \quad N_{1n} = 1.1 \times N_1 = 39.5 \text{ kN}$$

Solution 2: (Feng and Young [7.23])

$$\beta = b_1/b_0 = 40.3/80.6 = 0.50 < 0.70 \rightarrow \text{Chord face failure}$$

Equation (8.12) proposed by Feng and Young [7.23] is used as follows:

$$N_{1np} = \alpha_A N_{1n} = \alpha_A N_1 \times 1.1 = \alpha_A \frac{f_{y0}t_0^2}{(1-\beta)\sin\theta_1}\left[\frac{2\eta}{\sin\theta_1} + 4(1-\beta)^{0.5}\right]f(n) \times 1.1$$

where

$$\alpha_A = 1 - \frac{b_0}{100 t_0} = 1 - \frac{80.6}{100 \times 2.96} = 0.73$$

$$\eta = h_1/b_0 = 40.1/80.6 = 0.50, \quad f(n) = 1 - \frac{0.1n}{\beta}$$

in which

$$n = \frac{N_p}{A_0 f_{y0}} = 0, \quad \text{thus } f(n) = 1.0$$

Therefore, $N_{1np} = 28.7$ kN.

Comparison

The joint strength of stainless steel tubular T-joint of TD-C160 × 80 × 3-B40 × 40 × 2 was obtained from the experimental investigation described in Chapter 7 and presented in Ref. [7.20] as $N_f = 30.8$ kN.

For design formulae given in the CIDECT Design Guide No. 3 [8.3],

$$\frac{N_{1n} - N_f}{N_{1n}} \times 100\% = \frac{39.5 - 30.8}{39.5} \times 100\% = 22\%$$

Thus, the joint strength calculated using the design formulae given in the CIDECT Design Guide No. 3 [8.3] is larger than that obtained from the experimental investigation by 22%.

For design formula proposed by Feng and Young [7.23],

$$\frac{N_{1np} - N_f}{N_{1np}} \times 100\% = \frac{28.7 - 30.8}{28.7} \times 100\% = -7\%$$

Thus, the joint strength calculated using the design formula proposed by Feng and Young [7.23] is smaller than that obtained from the experimental investigation by 7%.

Therefore, the design formulae proposed by Feng and Young [7.23] are generally much more accurate than those given in the CIDECT Design Guide No. 3 [8.3] for stainless steel tubular T-joint.

Design Example 2

Determine the joint strength of stainless steel tubular T-joint of TD-C50 × 50 × 1.5-B50 × 50 × 1.5. Given that

SHS chord: 50 × 50 × 1.5 ($h_0 = 50.5$ mm, $b_0 = 50.2$ mm, $t_0 = 1.56$ mm, $r_0 = 1.50$ mm, $R_0 = 3.06$ mm, $A_0 = 298$ mm^2)

SHS brace: 50 × 50 × 1.5 ($h_1 = 50.5$ mm, $b_1 = 50.2$ mm, $t_1 = 1.57$ mm, $r_1 = 1.50$ mm, $R_1 = 3.07$ mm, $A_1 = 300$ mm^2)

$E_0 = E_1 = 200$ GPa, $f_{y0} = f_{y1} = 622$ MPa, $\theta_1 = 90°$, and $\sin \theta_1 = 1.0$.

Solution 1: (CIDECT No. 3 [8.3])

$$\beta = b_1/b_0 = 50.2/50.2 = 1.0 \rightarrow \text{Chord side wall failure}$$

Equation (8.8) given in the CIDECT Design Guide No. 3 [8.3] is used as follows:

$$N_1 = \frac{f_k t_0}{\sin \theta_1} \left[\frac{2h_1}{\sin \theta_1} + 10 t_0 \right]$$

where

$$f_k = \chi f_{y0} \sin \theta_1$$

in which

$$\chi = \frac{1}{\Phi + \sqrt{\Phi^2 - \overline{\lambda}^2}}, \quad \Phi = 0.5 \left[1 + \alpha(\overline{\lambda} - 0.2) + \overline{\lambda}^2 \right]$$

$$\overline{\lambda} = 3.46 \frac{((h_0/t_0) - 2)\sqrt{1/\sin \theta_1}}{\pi \sqrt{E_0/f_{y0}}}, \quad \alpha = 0.21$$

Therefore, $N_1 = 28.6$ kN, $N_{1n} = 1.1 \times N_1 = 31.5$ kN.

Solution 2: (Feng and Young [7.23])

$$\beta = b_1/b_0 = 50.2/50.2 = 1.0 > 0.85 \rightarrow \text{Chord side wall failure}$$

Equation (8.17) proposed by Feng and Young [7.23] is used as follows:

$$N_{1np} = \alpha_B N_{1n} = \alpha_B N_1 \times 1.1 = \alpha_B \frac{f_k t_0}{\sin \theta_1} \left[\frac{2h_1}{\sin \theta_1} + 10 t_0 \right] \times 1.1$$

where

$$\alpha_B = \frac{2}{25} \left(\frac{h_0}{t_0} - 1 \right) = \frac{2}{25} \times \left(\frac{50.5}{1.56} - 1 \right) = 2.51, \quad f_k = \chi f_{y0} \sin \theta_1$$

in which

$$\chi = \frac{1}{\Phi + \sqrt{\Phi^2 - \overline{\lambda}^2}}, \quad \Phi = 0.5 \left[1 + \alpha(\overline{\lambda} - 0.2) + \overline{\lambda}^2 \right]$$

$$\overline{\lambda} = 3.46 \frac{((h_0/t_0) - 2)\sqrt{1/\sin \theta_1}}{\pi \sqrt{E_0/f_{y0}}}, \quad \alpha = 0.21$$

Therefore, $N_{1np} = 79.0$ kN.

Comparison

The joint strength of stainless steel tubular T-joint of TD-C50 × 50 × 1.5-B50 × 50 × 1.5 was obtained from the experimental investigation described in Chapter 7 and presented in Ref. [7.20] as $N_f = 66.4$ kN.

For design formulae given in the CIDECT Design Guide No. 3 [8.3],

$$\frac{N_{1n} - N_f}{N_{1n}} \times 100\% = \frac{31.5 - 66.4}{31.5} \times 100\% = -111\%$$

Thus, the joint strength calculated using the design formulae given in the CIDECT Design Guide No. 3 [8.3] is smaller than that obtained from the experimental investigation by 111%.

For design formulae proposed by Feng and Young [7.23],

$$\frac{N_{1np} - N_f}{N_{1np}} \times 100\% = \frac{79.0 - 66.4}{79.0} \times 100\% = 16\%$$

Thus, the joint strength calculated using the design formulae proposed by Feng and Young [7.23] is larger than that obtained from the experimental investigation by 16%.

Therefore, the design formulae proposed by Feng and Young [7.23] are generally much more accurate than those given in the CIDECT Design Guide No. 3 [8.3] for stainless steel tubular T-joint.

Design Example 3

Determine the joint strength of stainless steel tubular T-joint of TH-C110 × 200 × 4-B150 × 150 × 6. Given that

RHS   chord: 110 × 200 × 4   ($h_0 = 109.2$ mm, $b_0 = 197.4$ mm, $t_0 = 4.05$ mm, $r_0 = 8.50$ mm, $R_0 = 12.55$ mm, $A_0 = 2345$ mm^2)

SHS   brace: 150 × 150 × 6   ($h_1 = 150.2$ mm, $b_1 = 150.2$ mm, $t_1 = 5.84$ mm, $r_1 = 6.00$ mm, $R_1 = 11.84$ mm, $A_1 = 3283$ mm^2)

$E_0 = 200$ GPa, $E_1 = 194$ GPa, $f_{y0} = 503$ MPa, $f_{y1} = 497$ MPa, $\theta_1 = 90°$, and $\sin\theta_1 = 1.0$.

Solution 1: (CIDECT No. 3 [8.3])

$$\beta = b_1/b_0 = 150.2/197.4 = 0.76 < 0.85 \rightarrow \text{Chord face failure}$$

Equation (8.1) given in the CIDECT Design Guide No. 3 [8.3] is used as follows:

$$N_1 = \frac{f_{y0}t_0^2}{(1-\beta)\sin\theta_1}\left[\frac{2\eta}{\sin\theta_1} + 4(1-\beta)^{0.5}\right]f(n)$$

where

$$\eta = h_1/b_0 = 150.2/197.4 = 0.76, \quad f(n) = 1.3 - \frac{0.4n}{\beta} \le 1.0$$

In which

$$n = \frac{N_p}{A_0 f_{y0}} = 0, \quad \text{thus } f(n) = 1.0$$

Therefore, $N_1 = 120.0$ kN, $N_{1n} = 1.1 \times N_1 = 132.0$ kN.

Solution 2: (Feng and Young [7.23])

$\beta = b_1/b_0 = 150.2/197.4 = 0.76 \rightarrow$ Combined chord face failure and chord sidewall failure

Equation (8.20) proposed by Feng and Young [7.23] is used as follows:

$$N_{1np} = \alpha_{A+B}N_{1n} = \alpha_{A+B}N_1 \times 1.1 = \alpha_{A+B}\frac{f_{y0}t_0^2}{(1-\beta)\sin\theta_1}\left[\frac{2\eta}{\sin\theta_1} + 4(1-\beta)^{0.5}\right]f(n) \times 1.1$$

where

$$\alpha_{A+B} = 1 + \frac{3b_0}{1000t_0} = 1 + \frac{3 \times 197.4}{1000 \times 4.05} = 1.15$$

$$\eta = h_1/b_0 = 150.2/197.4 = 0.76, \quad f(n) = 1 - \frac{0.1n}{\beta}$$

in which

$$n = \frac{N_p}{A_0 f_{y0}} = 0, \quad \text{thus } f(n) = 1.0$$

Therefore, $N_{1np} = 151.3$ kN.

Comparison

The joint strength of stainless steel tubular T-joint of TH-C110 × 200 × 4-B150 × 150 × 6 was obtained from the experimental investigation described in Chapter 7 and presented in Ref. [7.20] as $N_f = 145.4$ kN.

For design formulae given in the CIDECT Design Guide No. 3 [8.3],

$$\frac{N_{1n} - N_f}{N_{1n}} \times 100\% = \frac{132.0 - 145.4}{132.0} \times 100\% = -10\%$$

Thus, the joint strength calculated using the design formulae given in the CIDECT Design Guide No. 3 [8.3] is smaller than that obtained from the experimental investigation by 10%.

For design formulae proposed by Feng and Young [7.23],

$$\frac{N_{1np} - N_f}{N_{1np}} \times 100\% = \frac{151.3 - 145.4}{151.3} \times 100\% = 4\%$$

Thus, the joint strength calculated using the design formulae proposed by Feng and Young [7.23] is larger than that obtained from the experimental investigation by 4%.

Therefore, the design formulae proposed by Feng and Young [7.23] are generally much more accurate than those given in the CIDECT Design Guide No. 3 [8.3] for stainless steel tubular T-joint.

8.5.3 Stainless Steel Tubular X-Joint Without Chord Preload

●●●
──

Design Example 4

Determine the joint strength of stainless steel tubular X-joint without chord preload of XD-C140 × 80 × 3-B40 × 40 × 2-P0. Given that

RHS chord: 140 × 80 × 3 ($h_0 = 140.2$ mm, $b_0 = 80.2$ mm, $t_0 = 3.33$ mm, $r_0 = 6.50$ mm, $R_0 = 9.83$ mm, $A_0 = 1377$ mm^2)

SHS brace: 40 × 40 × 2 ($h_1 = 39.9$ mm, $b_1 = 40.3$ mm, $t_1 = 1.96$ mm, $r_1 = 2.00$ mm, $R_1 = 3.96$ mm, $A_1 = 289$ mm^2)

$E_0 = 212$ GPa, $E_1 = 216$ GPa, $f_{y0} = 486$ MPa, $f_{y1} = 707$ MPa, $\theta_1 = 90°$, and $\sin\theta_1 = 1.0$.

Solution 1: (CIDECT No. 3 [8.3])

$$\beta = b_1/b_0 = 40.3/80.2 = 0.50 < 0.85 \rightarrow \text{Chord face failure}$$

Equation (8.1) given in the CIDECT Design Guide No. 3 [8.3] is used as follows:

$$N_1 = \frac{f_{y0}t_0^2}{(1-\beta)\sin\theta_1}\left[\frac{2\eta}{\sin\theta_1} + 4(1-\beta)^{0.5}\right]f(n)$$

where

$$\eta = h_1/b_0 = 39.9/80.2 = 0.50, \quad f(n) = 1.3 - \frac{0.4n}{\beta} \leq 1.0$$

in which

$$n = \frac{N_p}{A_0f_{y0}} = 0, \quad \text{thus } f(n) = 1.0$$

Therefore, $N_1 = 41.4$ kN, $N_{1n} = 1.1 \times N_1 = 45.5$ kN.

Solution 2: (Feng and Young [7.23])

$$\beta = b_1/b_0 = 40.3/80.2 = 0.50 < 0.70 \rightarrow \text{Chord face failure}$$

Equation (8.12) proposed by Feng and Young [7.23] is used as follows:

$$N_{1np} = \alpha_A N_{1n} = \alpha_A N_1 \times 1.1 = \alpha_A \frac{f_{y0}t_0^2}{(1-\beta)\sin\theta_1}\left[\frac{2\eta}{\sin\theta_1} + 4(1-\beta)^{0.5}\right]f(n) \times 1.1$$

where

$$\alpha_A = 1 - \frac{b_0}{100t_0} = 1 - \frac{80.2}{100 \times 3.33} = 0.76$$

$$\eta = h_1/b_0 = 39.9/80.2 = 0.50, \quad f(n) = 1 - \frac{0.1n}{\beta}$$

in which

$$n = \frac{N_p}{A_0f_{y0}} = 0, \quad \text{thus } f(n) = 1.0$$

Therefore, $N_{1np} = 34.5$ kN.

Comparison

The joint strength of stainless steel tubular X-joint without chord preload of XD-C140 × 80 × 3-B40 × 40 × 2-P0 was obtained from the experimental investigation described in Chapter 7 and presented in Ref. [7.21] as $N_f = 25.5$ kN.

For design formulae given in the CIDECT Design Guide No. 3 [8.3],

$$\frac{N_{1n} - N_f}{N_{1n}} \times 100\% = \frac{45.5 - 25.5}{45.5} \times 100\% = 44\%$$

Thus, the joint strength calculated using the design formulae given in the CIDECT Design Guide No. 3 [8.3] is larger than that obtained from the experimental investigation by 44%.

For design formulae proposed by Feng and Young [7.23],

$$\frac{N_{1np} - N_f}{N_{1np}} \times 100\% = \frac{34.5 - 25.5}{34.5} \times 100\% = 26\%$$

Thus, the joint strength calculated using the design formulae proposed by Feng and Young [7.23] is larger than that obtained from the experimental investigation by 26%.

Therefore, the design formulae proposed by Feng and Young [7.23] are generally much more accurate than those given in the CIDECT Design Guide No. 3 [8.3] for stainless steel tubular X-joint.

Design Example 5

Determine the joint strength of stainless steel tubular X-joint without chord preload of XH-C150 × 150 × 6-B150 × 150 × 6-P0. Given that

SHS chord: 150 × 150 × 6 ($h_0 = 150.3$ mm, $b_0 = 150.5$ mm, $t_0 = 5.75$ mm, $r_0 = 6.00$ mm, $R_0 = 11.75$ mm, $A_0 = 3239$ mm^2)

SHS brace: 150 × 150 × 6 ($h_1 = 150.3$ mm, $b_1 = 150.3$ mm, $t_1 = 5.84$ mm, $r_1 = 6.00$ mm, $R_1 = 11.84$ mm, $A_1 = 3285$ mm^2)

$E_0 = E_1 = 194$ GPa, $f_{y0} = f_{y1} = 497$ MPa, $\theta_1 = 90°$, and $\sin \theta_1 = 1.0$.

Solution 1: (CIDECT No. 3 [8.3])

$$\beta = b_1/b_0 = 150.3/150.5 = 1.0 \rightarrow \text{Chord sidewall failure}$$

Equation (8.8) given in the CIDECT Design Guide No. 3 [8.3] is used as follows:

$$N_1 = \frac{f_k t_0}{\sin \theta_1} \left[\frac{2h_1}{\sin \theta_1} + 10t_0 \right]$$

where

$$f_k = 0.8\chi f_{y0} \sin \theta_1$$

in which

$$\chi = \frac{1}{\Phi + \sqrt{\Phi^2 - \overline{\lambda}^2}}, \quad \Phi = 0.5 \left[1 + \alpha(\overline{\lambda} - 0.2) + \overline{\lambda}^2 \right]$$

$$\overline{\lambda} = 3.46 \frac{((h_0/t_0) - 2)\sqrt{1/\sin \theta_1}}{\pi \sqrt{E_0/f_{y0}}}, \quad \alpha = 0.21$$

Therefore, $N_1 = 364.5$ kN, $N_{1n} = 1.1 \times N_1 = 400.9$ kN.

Solution 2: (Feng and Young [7.23])

$$\beta = b_1/b_0 = 150.3/150.5 = 1.0 > 0.85 \rightarrow \text{Chord sidewall failure}$$

Equation (8.17) proposed by Feng and Young [7.23] is used as follows:

$$N_{1np} = \alpha_B N_{1n} = \alpha_B N_1 \times 1.1 = \alpha_B \frac{f_k t_0}{\sin \theta_1}\left[\frac{2h_1}{\sin \theta_1} + 10t_0\right] \times 1.1$$

where

$$\alpha_B = \frac{2}{25}\left(\frac{h_0}{t_0} - 1\right) = \frac{2}{25}\times\left(\frac{150.3}{5.75} - 1\right) = 2.01, \quad f_k = 0.8\chi f_{y0}\sin\theta_1$$

in which

$$\chi = \frac{1}{\Phi + \sqrt{\Phi^2 - \overline{\lambda}^2}}, \quad \Phi = 0.5\left[1 + \alpha(\overline{\lambda} - 0.2) + \overline{\lambda}^2\right]$$

$$\overline{\lambda} = 3.46\,\frac{((h_0/t_0) - 2)\sqrt{1/\sin\theta_1}}{\pi\sqrt{E_0/f_{y0}}}, \quad \alpha = 0.21$$

Therefore, $N_{1np} = 806.2$ kN.

Comparison

The joint strength of stainless steel tubular X-joint without chord preload of XH-C150 × 150 × 6-B150 × 150 × 6-P0 was obtained from the experimental investigation described in Chapter 7 and presented in Ref. [7.21] as $N_f = 898.2$ kN.

For design formulae given in the CIDECT Design Guide No. 3 [8.3],

$$\frac{N_{1n} - N_f}{N_{1n}} \times 100\% = \frac{400.9 - 898.2}{400.9} \times 100\% = -124\%$$

Thus, the joint strength calculated using the design formulae given in the CIDECT Design Guide No. 3 [8.3] is smaller than that obtained from the experimental investigation by 124%.

For design formulae proposed by Feng and Young [7.23],

$$\frac{N_{1np} - N_f}{N_{1np}} \times 100\% = \frac{806.2 - 898.2}{806.2} \times 100\% = -11\%$$

Thus, the joint strength calculated using the design formulae proposed by Feng and Young [7.23] is smaller than that obtained from the experimental investigation by 11%.

Therefore, the design formulae proposed by Feng and Young [7.23] are generally much more accurate than those given in the CIDECT Design Guide No. 3 [8.3] for stainless steel tubular X-joint.

Design Example 6

Determine the joint strength of stainless steel tubular X-joint without chord preload of XH-C110 × 200 × 4-B150 × 150 × 6-P0. Given that

RHS   chord: 110 × 200 × 4   ($h_0 = 110.3$ mm, $b_0 = 196.3$ mm, $t_0 = 3.98$ mm,
$r_0 = 8.50$ mm, $R_0 = 12.48$ mm, $A_0 = 2305$ mm^2)
SHS   brace: 150 × 150 × 6   ($h_1 = 150.3$ mm, $b_1 = 150.4$ mm, $t_1 = 5.82$ mm,
$r_1 = 6.00$ mm, $R_1 = 11.82$ mm, $A_1 = 3276$ mm^2)
$E_0 = 200$ GPa, $E_1 = 194$ GPa, $f_{y0} = 503$ MPa, $f_{y1} = 497$ MPa, $\theta_1 = 90°$, and sin $\theta_1 = 1.0$.

Solution 1: (CIDECT No. 3 [8.3])

$$\beta = b_1/b_0 = 150.4/196.3 = 0.77 < 0.85 \rightarrow \text{Chord face failure}$$

Equation (8.1) given in the CIDECT Design Guide No. 3 [8.3] is used as follows:

$$N_1 = \frac{f_{y0}t_0^2}{(1-\beta)\sin\theta_1}\left[\frac{2\eta}{\sin\theta_1} + 4(1-\beta)^{0.5}\right]f(n)$$

where

$$\eta = h_1/b_0 = 150.3/196.3 = 0.77, \quad f(n) = 1.3 - \frac{0.4n}{\beta} \leq 1.0$$

in which

$$n = \frac{N_p}{A_0 f_{y0}} = 0, \quad \text{thus } f(n) = 1.0$$

Therefore, $N_1 = 118.1$ kN, $N_{1n} = 1.1 \times N_1 = 129.9$ kN.

Solution 2: (Feng and Young [7.23])

$$\beta = b_1/b_0 = 150.4/196.3 = 0.77 \rightarrow \text{Combined chord face failure and chord sidewall failure}$$

Equation (8.20) proposed by Feng and Young [7.23] is used as follows:

$$N_{1np} = \alpha_{A+B}N_{1n} = \alpha_{A+B}N_1 \times 1.1 = \alpha_{A+B}\frac{f_{y0}t_0^2}{(1-\beta)\sin\theta_1}\left[\frac{2\eta}{\sin\theta_1} + 4(1-\beta)^{0.5}\right]f(n) \times 1.1$$

where

$$\alpha_{A+B} = 1 + \frac{3b_0}{1000t_0} = 1 + \frac{3 \times 196.3}{1000 \times 3.98} = 1.15$$

$$\eta = h_1/b_0 = 150.3/196.3 = 0.77, \quad f(n) = 1 - \frac{0.1n}{\beta}$$

in which

$$n = \frac{N_p}{A_0 f_{y0}} = 0, \quad \text{thus } f(n) = 1.0$$

Therefore, $N_{1np} = 149.1$ kN.

Comparison

The joint strength of stainless steel tubular X-joint without chord preload of XH-C110 × 200 × 4-B150 × 150 × 6-P0 was obtained from the experimental investigation described in Chapter 7 and presented in Ref. [7.21] as $N_f = 170.1$ kN.

For design formulae given in the CIDECT Design Guide No. 3 [8.3],

$$\frac{N_{1n} - N_f}{N_{1n}} \times 100\% = \frac{129.9 - 170.1}{129.9} \times 100\% = -31\%$$

Thus, the joint strength calculated using the design formulae given in the CIDECT Design Guide No. 3 [8.3] is smaller than that obtained from the experimental investigation by 31%.

For design formulae proposed by Feng and Young [7.23],

$$\frac{N_{1np} - N_f}{N_{1np}} \times 100\% = \frac{149.1 - 170.1}{149.1} \times 100\% = -14\%$$

Thus, the joint strength calculated using the design formulae proposed by Feng and Young [7.23] is smaller than that obtained from the experimental investigation by 14%.

Therefore, the design formulae proposed by Feng and Young [7.23] are generally much more accurate than those given in the CIDECT Design Guide No. 3 [8.3] for stainless steel tubular X-joint.

8.5.4 Stainless Steel Tubular X-joint with Chord Preload

Design Example 7

Determine the joint strength of stainless steel tubular X-joint with chord preload of XD-C140 × 80 × 3-B40 × 40 × 2-P0.3. Given that

RHS chord: 140 × 80 × 3 ($h_0 = 140.1$ mm, $b_0 = 80.1$ mm, $t_0 = 3.10$ mm, $r_0 = 6.50$ mm, $R_0 = 9.60$ mm, $A_0 = 1284$ mm^2)

SHS brace: 40 × 40 × 2 ($h_1 = 39.9$ mm, $b_1 = 40.3$ mm, $t_1 = 1.91$ mm, $r_1 = 2.00$ mm, $R_1 = 3.91$ mm, $A_1 = 282$ mm^2)

$E_0 = 212$ GPa, $E_1 = 216$ GPa, $f_{y0} = 486$ MPa, $f_{y1} = 707$ MPa, $N_p = 187.2$ kN, $\theta_1 = 90°$, and $\sin \theta_1 = 1.0$.

Solution 1: (CIDECT No. 3 [8.3])

$$\beta = b_1/b_0 = 40.3/80.1 = 0.50 < 0.85 \rightarrow \text{Chord face failure}$$

Equation (8.1) given in the CIDECT Design Guide No. 3 [8.3] is used as follows:

$$N_1 = \frac{f_{y0}t_0^2}{(1 - \beta)\sin \theta_1}\left[\frac{2\eta}{\sin \theta_1} + 4(1 - \beta)^{0.5}\right]f(n)$$

where

$$\eta = h_1/b_0 = 39.9/80.1 = 0.50, \quad f(n) = 1.3 - \frac{0.4n}{\beta} \leq 1.0$$

in which

$$n = \frac{N_p}{A_0 f_{y0}} = \frac{187.2 \times 1000}{1284 \times 486} = 0.3$$

Thus,

$$f(n) = 1.3 - \frac{0.4 \times 0.3}{0.50} = 0.24$$

Therefore, $N_1 = 35.9$ kN, $N_{1n} = 1.1 \times N_1 = 39.5$ kN.

Solution 2: (Feng and Young [7.23])

$$\beta = b_1/b_0 = 40.3/80.1 = 0.50 < 0.70 \rightarrow \text{Chord face failure}$$

Equation (8.12) proposed by Feng and Young [7.23] is used as follows:

$$N_{1np} = \alpha_A N_{1n} = \alpha_A N_1 \times 1.1 = \alpha_A \frac{f_{y0} t_0^2}{(1-\beta)\sin\theta_1}\left[\frac{2\eta}{\sin\theta_1} + 4(1-\beta)^{0.5}\right] f(n) \times 1.1$$

where

$$\alpha_A = 1 - \frac{b_0}{100 t_0} = 1 - \frac{80.1}{100 \times 3.10} = 0.74$$

$$\eta = h_1/b_0 = 39.9/80.1 = 0.50, \quad f(n) = 1 - \frac{0.1n}{\beta}$$

in which

$$n = \frac{N_p}{A_0 f_{y0}} = \frac{187.2 \times 1000}{1284 \times 486} = 0.3$$

Thus,

$$f(n) = 1 - \frac{0.1 \times 0.3}{0.50} = 0.94$$

Therefore, $N_{1np} = 27.6$ kN.

Comparison

The joint strength of stainless steel tubular X-joint with chord preload of XD-C140 × 80 × 3-B40 × 40 × 2-P0.3 was obtained from the experimental investigation described in Chapter 7 and presented in Ref. [7.21] as $N_f = 25.1$ kN.

For design formulae given in the CIDECT Design Guide No. 3 [8.3]

$$\frac{N_{1n} - N_f}{N_{1n}} \times 100\% = \frac{39.5 - 25.1}{39.5} \times 100\% = 36\%$$

Thus, the joint strength calculated using the design formulae given in the CIDECT Design Guide No. 3 [8.3] is larger than that obtained from the experimental investigation by 36%.

For design formulae proposed by Feng and Young [7.23],

$$\frac{N_{1np} - N_f}{N_{1np}} \times 100\% = \frac{27.6 - 25.1}{27.6} \times 100\% = 9\%$$

Thus, the joint strength calculated using the design formulae proposed by Feng and Young [7.23] is larger than that obtained from the experimental investigation by 9%.

Therefore, the design formulae proposed by Feng and Young [7.23] are generally much more accurate than those given in the CIDECT Design Guide No. 3 [8.3] for stainless steel tubular X-joint with chord preload.

Design Example 8

Determine the joint strength of stainless steel tubular X-joint with chord preload of XH-C150 × 150 × 6-B150 × 150 × 6-P0.1. Given that

SHS   chord: 150 × 150 × 6   ($h_0 = 150.3$ mm, $b_0 = 150.2$ mm, $t_0 = 5.85$ mm, $r_0 = 6.00$ mm, $R_0 = 11.85$ mm, $A_0 = 3289$ mm^2)

SHS   brace: 150 × 150 × 6   ($h_1 = 150.3$ mm, $b_1 = 150.2$ mm, $t_1 = 5.85$ mm, $r_1 = 6.00$ mm, $R_1 = 11.85$ mm, $A_1 = 3289$ mm^2)

$E_0 = E_1 = 194$ GPa, $f_{y0} = f_{y1} = 497$ MPa, $N_p = 163.5$ kN, $\theta_1 = 90°$, and $\sin \theta_1 = 1.0$.

Solution 1: (CIDECT No. 3 [8.3])

$$\beta = b_1/b_0 = 150.2/150.2 = 1.0 \rightarrow \text{Chord side wall failure}$$

Equation (8.8) given in the CIDECT Design Guide No. 3 [8.3] is used as follows:

$$N_1 = \frac{f_k t_0}{\sin \theta_1}\left[\frac{2h_1}{\sin \theta_1} + 10t_0\right]$$

where

$$f_k = 0.8\chi f_{y0}\sin \theta_1$$

in which

$$\chi = \frac{1}{\Phi + \sqrt{\Phi^2 - \overline{\lambda}^2}}, \quad \Phi = 0.5\left[1 + \alpha(\overline{\lambda} - 0.2) + \overline{\lambda}^2\right]$$

$$\overline{\lambda} = 3.46 \frac{((h_0/t_0) - 2)\sqrt{1/\sin \theta_1}}{\pi\sqrt{E_0/f_{y0}}}, \quad \alpha = 0.21$$

Therefore, $N_1 = 383.0$ kN, $N_{1n} = 1.1 \times N_1 = 421.3$ kN.

Solution 2: (Feng and Young [7.23])

$$\beta = b_1/b_0 = 150.2/150.2 = 1.0 > 0.85 \rightarrow \text{Chord sidewall failure}$$

Equation (8.17) proposed by Feng and Young [7.23] is used as follows:

$$N_{1np} = \alpha_B N_{1n} = \alpha_B N_1 \times 1.1 = \alpha_B \frac{f_k t_0}{\sin \theta_1}\left[\frac{2h_1}{\sin \theta_1} + 10t_0\right] \times 1.1$$

where

$$\alpha_B = \frac{2}{25}\left(\frac{h_0}{t_0} - 1\right) = \frac{2}{25} \times \left(\frac{150.3}{5.85} - 1\right) = 1.98, \quad f_k = 0.8\chi f_{y0} \sin \theta_1$$

in which

$$\chi = \frac{1}{\Phi + \sqrt{\Phi^2 - \overline{\lambda}^2}}, \quad \Phi = 0.5\left[1 + \alpha(\overline{\lambda} - 0.2) + \overline{\lambda}^2\right]$$

$$\overline{\lambda} = 3.46\,\frac{((h_0/t_0) - 2)\sqrt{1/\sin\theta_1}}{\pi\sqrt{E_0/f_{y0}}}, \quad \alpha = 0.21$$

Therefore, $N_{1np} = 832.2$ kN.

Comparison

The joint strength of stainless steel tubular X-joint with chord preload of XH-C150 × 150 × 6-B150 × 150 × 6-P0.1 was obtained from the experimental investigation described in Chapter 7 and presented in Ref. [7.21] as $N_f = 871.1$ kN.

For design formulae given in the CIDECT Design Guide No. 3 [8.3],

$$\frac{N_{1n} - N_f}{N_{1n}} \times 100\% = \frac{421.3 - 871.1}{421.3} \times 100\% = -107\%$$

Thus, the joint strength calculated using the design formulae given in the CIDECT Design Guide No. 3 [8.3] is smaller than that obtained from the experimental investigation by 107%.

For design formulae proposed by Feng and Young [7.23],

$$\frac{N_{1np} - N_f}{N_{1np}} \times 100\% = \frac{832.2 - 871.1}{832.2} \times 100\% = -5\%$$

Thus, the joint strength calculated using the design formulae proposed by Feng and Young [7.23] is smaller than that obtained from the experimental investigation by 5%.

Therefore, the design formulae proposed by Feng and Young [7.23] are generally much more accurate than those given in the CIDECT Design Guide No. 3 [8.3] for stainless steel tubular X-joint with chord preload.

●●●

Design Example 9

Determine the joint strength of stainless steel tubular X-joint with chord preload of XH-C110 × 200 × 4-B150 × 150 × 6-P0.1. Given that

RHS chord: 110 × 200 × 4 ($h_0 = 109.0$ mm, $b_0 = 196.8$ mm, $t_0 = 3.97$ mm, $r_0 = 8.50$ mm, $R_0 = 12.47$ mm, $A_0 = 2294$ mm^2)

SHS brace: 150 × 150 × 6 ($h_1 = 150.3$ mm, $b_1 = 150.3$ mm, $t_1 = 5.87$ mm, $r_1 = 6.00$ mm, $R_1 = 11.87$ mm, $A_1 = 3301$ mm^2)

$E_0 = 200$ GPa, $E_1 = 194$ GPa, $f_{y0} = 503$ MPa, $f_{y1} = 497$ MPa, $N_p = 115.4$ kN, $\theta_1 = 90°$, and $\sin\theta_1 = 1.0$.

Solution 1: (CIDECT No. 3 [8.3])

$$\beta = b_1/b_0 = 150.3/196.8 = 0.76 < 0.85 \rightarrow \text{Chord face failure}$$

Equation (8.1) given in the CIDECT Design Guide No. 3 [8.3] is used as follows:

$$N_1 = \frac{f_{y0}t_0^2}{(1 - \beta)\sin\theta_1}\left[\frac{2\eta}{\sin\theta_1} + 4(1 - \beta)^{0.5}\right]f(n)$$

where

$$\eta = h_1/b_0 = 150.3/196.8 = 0.76, \quad f(n) = 1.3 - \frac{0.4n}{\beta} \leq 1.0$$

in which

$$n = \frac{N_p}{A_0 f_{y0}} = \frac{115.4 \times 1000}{2294 \times 503} = 0.1$$

Then,

$$f(n) = 1.3 - \frac{0.4 \times 0.1}{0.76} = 1.25 > 1.0, \quad \text{thus } f(n) = 1.0$$

Therefore, $N_1 = 116.5$ kN, $N_{1n} = 1.1 \times N_1 = 128.1$ kN.

Solution 2: (Feng and Young [7.23])

$$\beta = b_1/b_0 = 150.3/196.8 = 0.76 \rightarrow \text{Combined chord face failure and chord sidewall failure}$$

Equation (8.20) proposed by Feng and Young [7.23] is used as follows:

$$N_{1np} = \alpha_{A+B}N_{1n} = \alpha_{A+B}N_1 \times 1.1 = \alpha_{A+B}\frac{f_{y0}t_0^2}{(1 - \beta)\sin\theta_1}\left[\frac{2\eta}{\sin\theta_1} + 4(1 - \beta)^{0.5}\right]f(n) \times 1.1$$

where

$$\alpha_{A+B} = 1 + \frac{3b_0}{1000t_0} = 1 + \frac{3 \times 196.8}{1000 \times 3.97} = 1.15$$

$$\eta = h_1/b_0 = 150.3/196.8 = 0.76, \quad f(n) = 1 - \frac{0.1n}{\beta}$$

in which

$$n = \frac{N_p}{A_0 f_{y0}} = \frac{115.4 \times 1000}{2294 \times 503} = 0.1$$

Thus,

$$f(n) = 1 - \frac{0.1 \times 0.1}{0.76} = 0.99$$

Therefore, $N_{1np} = 145.3$ kN.

Comparison

The joint strength of stainless steel tubular X-joint with chord preload of XH-C110 × 200 × 4-B150 × 150 × 6-P0.1 was obtained from the experimental investigation described in Chapter 7 and presented in Ref. [7.21] as $N_f = 152.1$ kN.

For design formulae given in the CIDECT Design Guide No. 3 [8.3],

$$\frac{N_{1n} - N_f}{N_{1n}} \times 100\% = \frac{128.1 - 152.1}{128.1} \times 100\% = -19\%$$

Thus, the joint strength calculated using the design formulae given in the CIDECT Design Guide No. 3 [8.3] is smaller than that obtained from the experimental investigation by 19%.

For design formulae proposed by Feng and Young [7.23],

$$\frac{N_{1np} - N_f}{N_{1np}} \times 100\% = \frac{145.3 - 152.1}{145.3} \times 100\% = -5\%$$

Thus, the joint strength calculated using the design formulae proposed by Feng and Young [7.23] is smaller than that obtained from the experimental investigation by 5%.

Therefore, the design formulae proposed by Feng and Young [7.23] are generally much more accurate than those given in the CIDECT Design Guide No. 3 [8.3] for stainless steel tubular X-joint with chord preload.

8.6. SUMMARY

This chapter has shown that efficient finite element models not only are able to present the test results but are also able to generate more data outside the limits set in the test program. Not only that, efficient finite element models also assess the accuracy and validity of the design rules specified in current codes of practice and propose more accurate design guides and recommendations, which provide better understanding of the behavior for different metal structural members. The finite element analysis and design of cold-formed stainless steel tubular T- and X-joints were also described in this chapter. An extensive parametric study was performed using the verified finite element models developed in Chapter 7 to investigate the strength and behavior of cold-formed stainless steel tubular T- and X-joints of SHS and RHS. The effects of the critical geometric parameters and compressive chord preload on the structural performance of stainless steel tubular joints were evaluated in the parametric study, which were purposely designed beyond the validity range of those defined in the current design specifications.

The experimental and numerical results were compared with the joint strengths calculated using the current and proposed design rules for cold-formed stainless steel tubular T- and X-joints of SHS and RHS. It was shown from the comparison that the current design rules are generally quite unconservative for stainless steel tubular joints subjected to chord face failure, but it is quite conservative for specimens failed by chord side-wall. Furthermore, the current design rules are slightly conservative for stainless steel tubular joints subjected to combined chord face failure and chord sidewall failure, whereas it is generally appropriate for specimens failed by local buckling of brace. Lastly, some design examples are given to illustrate the design procedures of cold-formed stainless steel tubular T-, X-, and X-joint with chord preload subjected to different failure modes using the current and proposed design rules. It is shown from the comparison that the proposed design rules are generally much more accurate than the current design rules for cold-formed stainless steel tubular T-, X- and X-joint with chord preload.

REFERENCES

[8.1] Australian Institute of Steel Construction. Design capacity tables for structural steel hollow sections. Sydney, Australia: Australian Institute of Steel Construction, 1992.

[8.2] American Welding Society. Structural welding code-steel. Miami, FL: AWS D1.1/1.1M, 2004.

[8.3] Packer, J. A., Wardenier, J., Kurobane, Y., Dutta, D. and Yeomans, N. Design guide for rectangular hollow section (RHS) joints under predominantly static loading. Verlag TÜV Rheinland, Cologne, Germany: Comité International pour le Développement et l'Étude de la Construction Tubulaire (CIDECT), 1992.

[8.4] Eurocode 3. Design of steel structures—Part 1-8: Design of joints. European Committee for Standardization, CEN, Brussels, Belgium EN 1993-8, 2005

[8.5] International Institute of Welding. IIW static design procedure for welded hollow section joints-recommendations. IIW Doc. XV-E-04-301, 2004.

[8.6] Australian/New Zealand Standard. Cold-formed stainless steel structures. Standards Australia, Sydney, Australia AS/NZS 4673, 2001.

[8.7] Rasmussen, K. J. R. and Young, B. Tests of X- and K-joints in SHS stainless steel tubes. *Journal of Structural Engineering, ASCE*, 127(10), 1173–1182, 2001.

[8.8] Eurocode 3. Design of steel structures—Part 1-1: General rules and rules for buildings. European Committee for Standardization, CEN, Brussels, Belgium EN 1993-1, 2005.

INDEX

Note: Page numbers followed by "*f*" refer to figures respectively.

Printed and bound by CPI Group (UK) Ltd, Croydon, CR0 4YY

08/05/2025

01864907-0001